The Power of Twelve

Achieving 12-Strand DNA Consciousness

Anne Brewer

D0871081

SunShine Press Publications

SunShine Press Publications, Inc.,
PO Box 333
Hygiene, CO 80533

Cover design by Bob Schram of Bookends
Cover illustration by John Lewman

First Edition

Publisher's Cataloging-in-Publication Data

Brewer, Anne.
  The power of twelve: achieving 12-strand DNA
consciousness / Anne Brewer. — 1st ed.
    p. cm.
  Preassigned LCCN: 97-62466
  ISBN 1-888604-07-7

    1. Astral body. 2. DNA—Miscellanea. 3. New Age movement.
4. Self-realization. 5. Mental healing. 6. Spirit writings. I. Title

BF1389.A7B74 1998   133.9
                                           QBI98-6

Printed in the United States

5 4 3 2 1

Printed on recycled acid-free paper using soy ink

Table of Contents

Publisher's Note

The materials presented in this book include references to "channeled" material concerning spiritual and psychological matters which the author represents that she has received and translated to the best of her ability. However, according to the author, the translation of channeled information is subject to the inherent limitations of the English language, the written word, and the author's abilities, ego and belief system.

The materials presented and opinions expressed herein are not necessarily those of, or endorsed by, the Publisher.

This book deals with subject matter about energetic DNA. If you have questions or concerns about physical DNA, consult with your physician or health care provider.

Introduction

Seek your love now,
There is no past,
And tomorrow relies on the passions of today,
Forever.
—Robert Henion

Today, I experience expanded consciousness and greater power. I carry twelve strands of DNA in my astral body which plug into my physical body through my endocrine system. This additional DNA enables me to have access to twelve rather than two levels of information, resulting in the expansion of my consciousness. I make decisions that are no longer based in fear and guilt since both have been eliminated from my mental body, thus giving me a sense of greater power. I have an increased capacity for love and joy. I am a creator rather than a receiver. I feel a sense of purpose and fulfillment. I live in the now rather than projecting my experiences into the past or the future.

No one could have explained to me how it feels to be awakened after so many lifetimes of living in ignorance. My only point of context, until now, has been the tremendous sense of yearning for something I had lost but could not define. That yearning was the motivation that brought me to my current point of awareness. There are many paths available to each of us for reaching expanded consciousness. This is the path I have chosen for myself and, since it has significantly improved my life, it is the path I have chosen to share with others. I have a soul destiny contract to experience DNA recoding, bringing this particular method of retrieving twelve strands of DNA to other Earth beings who desire it. For those who resonate with this process, please join me on this journey. For those who do not, I share my story and hope that you learn something that might be helpful to you in your own process.

The transition to the fifth dimension is nearly upon us. Although we live in third dimensional bodies, we currently reside in the fourth dimension which is the realm of length, breadth, depth, and time for all animate beings, e.g., humans, plants, animals, and insects. We are currently transitioning to the fifth dimension which is the realm of length, breadth, depth, time, and spirituality for human beings. As we move closer and closer, we expand our self-conception to include our Divinity, encompassing us in new physical, mental, emotional, and spiritual aspects. We will recognize each other as we move toward the Light since we will finally be aligned with our Divineness through our new consciousness. We are a family of Light beings who originated from the single source of the Divine Creator.

In April, 1996, I began my journey to acquire twelve strands of DNA. Prior to that, I had discovered through reading channeled information that the Pleiadians, Kryon, Sananda, and many other non-physical beings frequently referred to the re-acquisition of our twelve strands of DNA in order to function as Spirit in physical form. This information created my exuberance for regaining my twelve strands.

The previous summer, I had read the first Kryon book which dealt with the subject of ascension to the fifth dimension and how it would probably be necessary to have different spirit guides to assist with the DNA acquisition process needed for ascending. Although I deeply desired to regain my DNA, I was not sure I was ready to relinquish my four guides who had helped me, protected me, and taught me. The Kryon book made it clear I would be requesting new guides to help me on my journey. Not knowing whether I was ready or not, I chose to acquire twelve strands of DNA. I then received the Kryon implant as described by Lee Carroll which canceled all of my remaining Earthly karma, a key criteria for beginning the recoding process.

It took a year of routine affirmations and spiritual growth to manifest my request to receive the ten missing strands. There are ancient prophecies for reference, but there are no guide books or medical dissertations to tell one how they will feel as their circuitry is reconnected and the energy starts to move along the paths that have been severed for eons. I decided to keep a record of the events that occurred from the day I was granted recoding. The intent of the journal was twofold, both for me to remember upon reflection what had transpired after I reached my destination, and to inform others what they might experience, should they choose this path.

In the beginning of this process, I did not fully realize I was the sole creator of my life experiences. I viewed my guides as the ones who were orchestrating my life and managing the recoding. I have since gained such understanding and empowerment that I now realize I am the one orchestrating my life from a physical and non-physical perspective. I was sent here to assist both the physical and non-physical world, and I can only do that as a self-empowered individual. My work consists of a co-creative effort set forth by my intention to reach expanded consciousness. I believe that Divine Creator energy resides in all of us, once we identify the Creator power within ourselves. The quality of the outcomes we experience from life are contingent on our understanding of ourselves as the Divine Creator and our retrieval of that power.

Love and peace on your journey,
Anne Brewer

A Note from the Non-incarnate Recoding Team

The following information on DNA recoding was made available from two non-physical entities; Anu, who is head of the Nibiruan Council, and Joysia, who is chief genetic engineer of the project. The Nibiruan Council is a group of non-incarnate beings who have karmic connections with Earth's evolution.

DNA recoding is the process by which an entity of human origin who has two strands of DNA re-acquires the twelve strands of DNA energetically that were originally his or her Divine right. The reconnection of this complete circuitry enables a soul to have access to twelve levels of spiritual, emotional, physical, and mental awareness and information versus the two levels afforded by two strands of DNA. Access to twelve levels results in expanded consciousness and greater power in third dimensional physical form which manifests in characteristics such as enhanced telepathic abilities like clairvoyance and clairaudience, memory recall of past, present, and future physical and non-physical existences, removal of fear and guilt from the mental body, and a complete balance of the endocrine system which is the receptor site for one's spiritual blueprint and the pathway to Divinity. Recoding is necessary to ascend. A human being must have twelve energetic or physical DNA strands to evolve to the next level, i.e., the fifth dimension. Ascension is the process of moving to the next dimensional level with full consciousness.

A Channeled Message from Anu:
Head of the Nibiruan Council

Greetings brothers and sisters of the Light. It is time for you to embrace a wonderful gift that is now yours. It is the opportunity to gain your full power through the realignment and fusion of your twelve strands of DNA. It is with great joy that I come to serve you by sharing the knowledge with you of this endeavor that you are now ready to embrace. I will give you a brief background on how this came to be.

I am of your parent race, the Pleiadians, and I am the one who was ruling the planet Nibiru, the artificial planet of the

Pleiades, when the two stranded DNA being was created. Nibiru was an explorer planet, a battle star, that was sent into the solar system and galaxy to protect the human colonies on all the planets and to battle the dark forces when they attempted to destroy these colonies. We came to your solar system when your planet was known as Tiamet, twice the size that it is today. Because the dark forces were overrunning this planet and destroying the evolving human civilization, we were given the task of destroying Tiamet to the point that Tiamet was broken in two. One part became the Earth and the other part the Milky Way Galaxy. This destroyed the basis of the dark ones, the Reptoids, who at that time were on the planet. Your planet then became known as Terra, a portion of Tiamet and one you call Earth. Earth and Terra are the same.

We, through millions of years with the Sirians and the Spiritual Hierarchies, made Terra the beautiful planet that you see today. It was also colonized. We came 480,000 of your years ago to your planet to mine gold and also to fulfill a contract with the Galactic Federation and the Spiritual Hierarchies to create a two-stranded DNA being for a particular group of souls to be able to inhabit your planet. This soul group had come into your Earth plane and had gotten trapped reincarnating into the animals and creating monstrosities of half-human/half-animal. These were beautiful souls who had made this choice but could not seem to move beyond it, and we were given the task of creating a vehicle that would give them the opportunity to slowly evolve. Naturally, we also saw an opportunity to use these souls as they evolved, employing them in the mines that were established on Terra at the time to extract gold for use on Nibiru. My son, Enki, and my daughter, Ninurshag, created the two-stranded beings for this group.

Although my children created a two-stranded being, they also included the additional ten strands carried by the Pleiadian race. This was done by artificially inseminating the Pleiadian women with sperm from the original Terra soul group. Two of the strands were coiled to create the necessary DNA helix to create the new being. Obviously, two strands were needed since this was the minimum number needed to create any animate being. The remaining ten strands were demagnetized with implants. You may wonder why twelve strands were placed in the physical being of these souls rather than confining it to two strands. Twelve strands were implanted in order to infuse the promise of the ultimate potential of this new soul group.

We, as the parent race, watched over this evolving being. Initially, the being was very limited as we used it to work the mines. It exhibited a herd behavior but could not propagate. Eventually, my children Enki and Ninurshag, felt compelled to upgrade the being in the form of Adamis, the story of your creation. This is the version of the vehicle you inhabit. We have been with you through this journey of all these thousands of years, sending beings from other places such as the Office of the Christos to expedite your growth. They would keep the Christ consciousness anchored as this two-stranded soul group continued to evolve.

Now, you are the result of many eons of evolution, and you are at the place where you are ready to own your full power. I, Anu, have come to share this knowledge with you and to let you know you are ready for that power. I am karmically connected because of my children and myself. I wish to serve you and be here with you as you regain this power that is rightfully yours. I am heading up the Nibiruan Council, working with the Galactic Federation, the different councils of the Federation, and with your Planetary Hierarchies to fulfill this wonderful endeavor. I wish you all the best in going through this process, and I am with you. I love you as any parent would love their children.

Go in peace and in the Light.
Anu

A Channeled Message from Joysia:
Chief Genetic Engineer

I am Joysia, a consultant of the Galactic Federation to the Nibiruan Council. I am the Chief Genetic Engineer in charge of the DNA recoding process for humans on Terra. It is with great pleasure that I come to serve you in this endeavor. Recoding is a wonderful right that you have earned, the ability to own your full power and be able to funnel it into your soul contract and complete your work while you are on Earth. This will give you the great joy, prosperity and happiness for which you long.

I would like to begin by giving some information about how this has occurred. During the Harmonic Convergence of 1987, we knew that we had moved forward in great leaps as far as being able to put this plan into practice. We knew, though, that it would take some time before the souls who had come together for that clearing would be able to clear enough density from their emotional and physical bodies so they would not experience intense pain from DNA recoding. On March 21, 1996, on your Spring Equinox, you were cleared for recoding. Now, there are those who we would not hold back who were recoded before this time, but there were very small groups of them. The mass plan was put into place on March 21 when enough of you were considered clear enough to begin recoding. At that time, we knew the percentage of success was so great that we could not fail because you had willingly worked on clearing your bodies through the different modalities of healing that you had learned and experienced and brought forward from past lives.

We knew that we needed to begin with those who originated from other planets and star systems—the starseeds. Because of their special forms of coding, this would enable them to release and clear much easier than the denser humans of that original Terra soul group for which the physical bodies were created in the first place. Therefore, we needed to get the starseeds who currently reside on Earth in third dimensional form to recode first because they would give us a great leap forward in establishing this new higher frequency for the planet. Once enough starseeds have recoded, they will stimulate others to respond. We knew there would be others of the Terra soul group who would be able to recode, but it would take them longer. We needed to initially work with the ones who could move the quickest. Eventually, some of those among the general population will desire to experience DNA recoding, for there are humans of this Terra soul group who need to ground the twelve stranded DNA frequency in the Earth plane.

There is a saying in your Bible that says that knowledge must go out to all people in all parts of the world and then the end would come. This must be fulfilled, and this is the knowledge of recoding. The end is the end of the era of darkness, and the beginning of an era where the veil has been dissolved and there will be no separation of dimensions. Recoding is the process through which a human with a two stranded DNA vehicle is able to realign and recapture the twelve strands. This is done through the astral body and filters into the physical body through the endocrine system. It is not done directly on the physical body because you would not wish to stay in it. You would leave it because it would be too difficult and too painful. As much clearing as you have done to your physical body, it is still too dense to hold twelve strands of DNA. Only a spirit body can hold twelve strands.

In order to recode, you must have the DNA implants removed that are demagnetizing or decoiling your twelve strands. You must give us permission to work on your astral body to reclaim your remaining ten strands. If you are a starseed, we have communications with your original soul group, and they decide which way to go. We also counsel with your High Self so you are always involved, even if it is on a subconscious level. We actually work on your astral body during your sleep state. The work filters down and is felt by the physical body just as the emotion of joy is felt in the physical body in the form of lightness and elation.

Through the fusion of the twelve strands, you will be able to do things with your mind that will affect your environment such as receiving energy from the universe more effectively. Also, you will be capable of receiving and directing energy through your hands for healing. These abilities will be felt as greater power, a release of fear, a release of guilt, which is all in the mind (or mental body). The power comes through the mind and is felt in the body as joy because there is no longer the implants of fear and guilt. Also, you feel much lighter because of the clearing of these implants. However, physiological fear to a physically threatening situation is always ready to spring forth at a moment's notice. Yet, you are still a third dimensional being. As the days unfold you will feel yourself shifting and changing because you have completed

DNA recoding, and you will become more fully seated into new abilities and new feelings. There are no words adequate to describe this feeling. It is otherworldly.

With great love and respect,
Joysia

Chapter One

April, 1996

Meeting Laramus and Devin

April 17, 1996 was the day I learned about recoding from several non-incarnate entities who would become a fundamental part of my everyday existence. It began when I visited an acquaintance who was called "N" by the entities she channeled, upon her invitation. She had been trance channeling a being named Devin for several months, and Devin had requested I attend a session with him. N and I were both curious about the request since we barely knew each other. We had met through a mutual friend, but we were not friends nor even casual acquaintances. I went to N's office on that Friday afternoon, bursting with curiosity about what Devin had to say. We dimmed the lights, shut the blinds, and settled back for a channeling session that would dramatically change my life. N put herself in a relaxed state and, almost immediately, Devin came through and introduced himself through N. I felt nervous in his presence, sensing his power and feeling his energy surge through the room. He said there was another entity wishing to come forward to speak to me, one who had been a dear friend of mine for a long time. I watched as Devin departed, and a new entity entered.

A different voice emerged as the newcomer introduced himself. He said he was a close friend who had lived many prior lives with me, both physical and nonphysical, and that he was happy to be with me at this time. It was strange to feel the love and closeness exuding from this entity of whom I had no conscious recall. I apologized for failing to recognize him or remember our past together and asked what I should call him. He said there was no word that came close to his real name but to use the name Laramus because the syllables sounded the tones that represented his energy.

He said he had "checked in" with me several times during my current life time but had not been very involved to date. However, he said he would now be available on a regular basis as my evolution had proceeded to the point where it was necessary for us to reconnect. He called himself my "genetic engineer." I immediately realized he was the "new guide" mentioned in the Kryon material.

Laramus asked if I was ready to accept the soul contract I had come to Earth to fulfill in preparation for receiving my twelve DNA strands. I said I was unfamiliar with the specifics of any contract although I knew my purpose involved helping the movement of the planet toward the fifth dimension. In fact, I was in the process of establishing a spiritual seminar business. He said he knew I was consciously unfamiliar with my contract, but he could not enlighten me at the time since I was not "ready" to receive the information. Again, Laramus asked if I was ready to accept my soul contract. Intuitively I knew I could accept this contract because I was a Light being who desired a better future for this planet. This gave me some assurance that my soul would only do work that aligned with my higher power and the universal Creator Truth. I also knew I was a member of several soul societies that were part of the Galactic Federation, an organization that was committed to improving the evolutionary cycle of this and other galaxies by moving souls toward spiritual enlightenment. I assumed I would be agreeing to a contract that enhanced the greater good of all.

Nonetheless, being trained in corporate business for fifteen years, I began negotiating. I asked if, by accepting my soul contract, I would be forced to radically change my life, relinquishing my loved ones for some spiritual quest or being asked to leave a flourishing market research business for a life devoted to some unknown contract. I was assured I lived in a free will zone, and I would always have choices in my life. I was told there would be change, but not more change than I could handle. I was told the choices I would need to make would come when I was ready for them although not all of them would be gentle. Finally, I was reminded that those who remained in my life were meant to be there and those who left were meant to proceed to something else. However, something in Laramus' manner made me wonder how this would affect my relationship with my partner, Jerry. I knew it would not have a negative influence on my relationship with my son, Drew, since there was an unbreakable bond between the two of us.

Jerry and I were already having difficulties. As I changed spiritually, the distance between us grew. We had known each other for twelve years. In fact, I had met him when he was married, working for him over a period of two years in New Jersey. This was prior to the company relocating me to Kansas City. We had established a strong friendship from the beginning, keeping in touch during the years following our working relationship when I resided in Kansas. Jerry divorced his wife in 1993 and came to Kansas City on business. During that trip, we met for dinner and a relationship kindled. Subsequently, we discovered our strong attraction was due to the fact that we were divine soul complements, which essentially meant our souls balanced each other like the Chinese yin and yang elements. We had many past lives together and were meant to walk a spiritual path together in this one. However, Jerry was having difficulty with this transition, preferring his old ways of thinking to my new ones, despite the fact his were no longer bringing him satisfaction. We loved each other deeply, but we were constantly struggling with our diverging philosophies. I wondered what my options for a partner were, given the divine soul complement status of our relationship. Was he my one and only partner? If accepting my contract pushed me further down my spiritual path, would it split me from the one who was meant to be with me?

I made one last effort to ensure stability in my life as I moved into this new world. After all, the promise of recapturing my twelve strands of DNA was enticing. I asked that the choices I needed to make be easy. Also, I requested that my transition be gentle, and

that this being named Laramus who professed to know me very
well did not abandon me as I made my way down this undefined
path. When I was reassured the transition would be as easy as
possible and that change would occur according to my needs, I
agreed to accept my soul contract.

I asked Laramus when I would receive my twelve strands of
DNA in order to better accomplish my goals. He said it was
necessary to formally ask for them, a process he called "DNA
recoding." I was surprised since I had been requesting twelve DNA
strands for the better part of a year. However, Laramus said there
was very specific wording that was needed to formally request this
form of recoding. Apparently, a soul must unequivocally accept its
soul contract in order to recode since there is hesitation to place
increased power in the hands of someone who may not be fully
committed to the Light energy. After I had agreed to accept my
soul contract, I was able to formally request recoding.

Next, Laramus directed me to transcribe his words and
proceeded to give me the exact terminology for my formal
recoding request which specifically addressed the Sirian/Pleiadian
Council. This wording is included in Appendix A: Summary of the
DNA Recoding Process, Phase I on page 154 which makes up the
first phase of recoding. After I wrote down the words, I asked
Laramus if I could read the request for recoding, and he invited
me to do so. I reread what I had written and was told to wait while
the Sirian/Pleiadian Council reviewed the request, since this was
the Council that was orchestrating this event. The information was
energetically sent before the Sirian/Pleiadian Council with permission being granted and recorded in the Akashic records.

This approval process took several moments during which
time I patiently awaited my outcome. Laramus communicated with
great pleasure the approval of my request. Tears were in my eyes
as I thanked Laramus for his assistance. I was touched by the
passing of this key event, having requested the return of my twelve
strands many times over the past several years. It was important
to regain the capabilities I had agreed to relinquish as a starseed
prior to coming to Earth as part of my soul growth experience.
But, I yearned for the return of what had once been an integral

part of me since I was convinced these abilities would assist me in fulfilling my planetary destiny. Laramus said he would now be close to me at all times, that I could consider him my personal genetic engineer who would help me through recoding. At that point, Laramus said goodbye and told me we would be communicating directly in the near future. Devin returned to speak briefly with me.

I asked Devin if I would be teaching recoding to individuals since my own guides had sent information to me to begin constructing a course on the "shift" into the fifth dimension. At this stage in my spiritual growth, I was still asking channeled entities for direction rather than relying on my intuition. Again, this is a habit that would change as a result of the Creator power I retrieved while recoding. Devin said to be patient, that the reason I had been given the formal terminology for requesting recoding was because I would be sharing it with others. Devin continued, "Don't you need to experience what you teach? Haven't you just experienced something? Why do you think we gave this to you?"

N returned from her deep trance, and I told her what had happened. We agreed to remain in touch since she, too, was undergoing recoding. I left feeling slightly dazed by all of the information I had just received. However, I also strongly resonated with it. Deep in my heart, I felt the appropriateness of the path that had been presented to me. I was excited by the prospects that awaited me as a fully conscious being.

Immediate Confirmation

I waited impatiently for two days for a new feeling—something—anything—to be different, to prove that I was actually, finally in the process of receiving my twelve strands. But, I felt like the same person. I hoped my guides would communicate with me during my sleep state, something they frequently did. I awoke in the middle of the night because my body vibrated up and down like a jack hammer was on my back. I was not frightened because I intuitively knew this had something to do with recoding. However, just to be sure, I called out to Laramus to intervene if this riveting energy was not for my higher good. When I called his name, it came out in a staccato tone, L-a-r-a-m-u-s, as if someone or thing was hammering my body. I laughed at how ridiculous I must look and sound, and immediately plunged into a deep sleep that I could not seem to resist.

In the morning, I awoke and asked my guides and Laramus to speak to me, explaining what had happened. I had recently begun automatic writing so I could access information from my guides by sitting down at my computer and listening to the words they put in my head. However, I wanted a simpler and more immediate approach to communication. As I lay there with my eyes closed, I "saw" a bolt of white light come toward my body. My entire body was physically jolted as the ball of light entered my body. The jolt was more startling and energizing than uncomfortable. I didn't know what it was but knew someone was trying to give me something. Later, I realized that Laramus was simply answering my request to feel something different. He sent me the ball of light to prove his presence. He showed me the immediacy with which my requests would be answered as I moved from the denser, slower manifestation of a two-stranded being to the lighter, faster manifestation of a recoded being. The energetic transmission was enough to satisfy my request for a signal that something new and different was happening to me.

Heart Healing in Preparation for Recoding

I visited Bobbie, my friend and intuitive counselor. As usual, I asked her to psychically scan my body for any distressed areas. Try as I might, I was not yet successful at keeping unwelcome energies out of my space. Bobbie helped keep my energy clean in support of my efforts to maintain optimum health.

Today, in scanning, Bobbie noticed a ball of light in the middle of my abdomen. She felt it was a positive energy, but she could not view it because it was enclosed in a metal sheathing, much like a missile casing. There were spider-like webs branching out from the sheathing, extending down to my ankles and up to my forehead. Bobbie told me I would need to request that the sheathing be removed before she could help me. I complied by asking my guides to remove the sheathing.

Bobbie began doing the energy work for which she was noted, and I focused my attention on bringing my guides and my High Self forward to assist. As she worked, I felt the presence of a deep purple light inside me. The light eventually shifted to a bright orange as I saw two wood nymphs with fluttering wings, square off, back to back, and then walk in opposite directions on either side of me. After several moments, Bobbie asked me what I had experienced, and I described it to her. She then told me what she had seen. She had encountered difficulty removing the sheathing and had asked for assistance. Several beings came forward, working hard to remove the metal. The white light turned into a bright orange light. At that point, Bobbie realized that the light behind the metal was my true essence which I had kept hidden for eons. She felt I was too vulnerable with it entirely exposed, and as she wondered what to do, two wood nymphs with wings emerged and stationed themselves at either side of me. They told her they were there to protect me when it was unwise for my essence to be fully present. She also watched as they wove this golden net around me, replacing the metal sheathing with this etheric web that was studded with jewels. She commented how beautiful the structure was although she was unsure of its purpose.

She then noticed my abdomen because it began protruding to the point of looking as if I were pregnant. Just as she began to worry how much larger I would "grow", an unusual being emerged from my womb. It was the same bright orange color of the ball of energy in my abdomen which now retained its normal shape. She asked the unusual looking entity if it was a positive being of Light and for my higher good. She asked the origin of the being and its purpose. The being said I had visited "a world on the edge of your universe" many hundreds of thousands of years ago, and had carried a seed housing this being in my system until such time as I would know to let it free. The being was to remain with me for a time, helping me to achieve my spiritual goals and then leave me to work on the Earth plane to help expand human consciousness.

As Bobbie relayed the information to me, we noted our matching vision of the color of the energy and the wood nymphs.

She referred several times to the being as "It," and then laughed, saying "It disliked being addressed in that manner." Although there were no Earth names close to its otherworldly language, it asked to be called Raphael since his energy was similar to that of Archangel Raphael's realm. He was here on Earth to bring the Raphaelite attributes of love and healing to those who needed it. Had Raphael been birthed sooner, he would not have been able to survive our dense energy field. However, the Earth had lightened sufficiently for him to emerge.

Bobbie commented as our session ended about how calm I remained upon hearing this information. She said the person she knew several years ago would probably have been very upset to hear she had "birthed" a strange looking being. Later, I would understand the extent of the healing energy Raphael brought to me as I opened my rejuvenated heart energy to my eternal lover, my life partner, Asalaine. My heart energy had been closed for the majority of my current lifetime based on a series of past lives that had turned me bitter toward my Earth experience. Raphael worked with me to resurrect the heart energy I had relinquished due to the painful emotions I had experienced, preparing me for the open heart I would require to connect with the man who was meant to be with me forever.

Emerging Psychic Energy

I was beginning to understand how recoding enables an individual to tap into his or her psychic ability by opening the all-seeing pipelines that have been blocked. Although some human beings already have intuitive power without ten additional DNA strands, this is unavailable to the mass population due to the fact that society trains us to honor the left rather than the right, or intuitive, side of our brain. Most of us are born with receptivity to being psychic but quickly lose that attitude as we learn to assess the world through five rather than six senses. By the time we are adults, it is very difficult to train our linear minds to reopen the intuitive channels that would enable us to be psychic.

However, recoding reconnects us with a psychic hookup. Although I had become increasingly reliant on my intuitive side and practiced automatic writing, I never considered myself truly psychic and was excited by the prospect of becoming "all-seeing." Then, I experienced my first psychic hit that would be a precursor of many more. My business associate had handled a two hundred piece mailing to my clients and prospects while I was on vacation. The brochure had been developed by a creative freelance artist so it was cute, fun and noticeable. A month had passed, and I was surprised that none of my clients had remarked about the brochure. I also had not received any returns which was highly unusual.

I called several of my regular clients and asked if they had received my mailing. No one remembered it. Next, I called my assistant at home and asked her if she remembered when and under what circumstances she had mailed it since no one had received it. She said she had mailed it when I was on vacation the previous month. As she responded, I clearly saw the boxes of mail in the trunk of her car. However, not being used to receiving visual transmissions, I dismissed it. The next morning, my assistant sheepishly entered my office. She reluctantly informed me that she really thought she had mailed the letters when speaking to me yesterday but later realized they were still in the trunk of her car.

I was amazed I had seen those envelopes so clearly and noted I would need to pay more attention to fleeting visuals because they were not as insignificant as I thought. I understood the implications of psychic power and why so few were afforded the ability to tap into this pool of information. If an individual was not balanced emotionally or felt a low level of responsibility toward others, this type of information could be used to weaken rather than strengthen. Ideally, psychic powers should be expanded among those who would use that ability to heal others or to help evolve the planet to a higher level of consciousness.

Discerning Truth

N called to see how I was doing with recoding, and I asked her to channel Devin for me to obtain a status report. She agreed, and Devin began by saying he was very pleased with the progress I was making in recoding. As I began asking questions about some trends I was seeing in my business, and some decisions I wished to have assistance making, Devin repeatedly reiterated that I already knew the answers to those questions. Devin felt I was asking him superfluous questions. He felt if I trusted the answers I was receiving internally, I would not feel the need to query him. He said my internal information was transmitted from the "purest" source since I had asked for recoding.

At the same time, I was frustrated when he put me off since I wanted information from a source who I perceived to have higher power. Now, I understand what he was trying to tell me during a stage in my life when I was still looking for direction from beings I thought were more evolved than myself. When we align with Creator Truth, we are connected to the best information available. Devin was trying to tell me I would make proper choices based on the guidance of my higher power, balanced with my own free will. I did not need to rely on the input of nonphysical beings who sounded all-knowing to formulate my life or provide me with direction. I needed to listen with my ears but heed with my heart, and I would find my own truth.

Devin used a wonderful analogy by comparing he and my spirit guides to shepherds. He said it was their job to gently guide me to the pastures that offered the highest nutrition. It was up to me to decide if I wanted to remain with the flock and partake of the nourishment or stray to other pastures. However, it was the quality of my choices that determined how quickly I moved through recoding toward expanded consciousness. He exited after saying there were many "old friends" around me that loved me very much and missed communicating with me. He said they were anxious for me to become an "open channel" in order to establish lines of communication so we could reconnect. As always, I felt frustration because I was not able to see or hear them and was anxious for those abilities to return. I asked, "When will I be able to channel?" He responded, "When you are ready, dear one."

Identifying Lahaina

I awoke suddenly at 3 a.m. to the sound of a loud voice. It resounded in my ears as it told me that my eternal name was

"Lahaina." Initially, I felt confused. It occurred to me this was the name of a town on the Hawaiian island of Maui that I had recently visited during my trip with Jerry. I wondered whether I was receiving my soul name or reliving the trip in a dream.

Jerry and I had gone to Maui to get married in September, but when we arrived the feelings of our spiritual incompatibility surfaced so strongly that we postponed the wedding. We knew we were going to be living together anyway since we were co-owners of a house so we enjoyed a vacation rather than honeymoon. Looking back, I understood that part of that trip to Hawaii was to establish knowledge about the area called Lahaina because it was an integral part of who I was. The name connected me to a past Polynesian lifetime with my soon-to-be-encountered true life partner, Asalaine. It would be one of the clues that brought us together. The name, Lahaina, also eventually identified my existence on the Nibiruan Council which I would soon discover.

Self-empowerment

I received an unanticipated call from my HMO informing me I had "questionable" lab results from a pap smear and would need to make an appointment for a kolposcopy and a possible biopsy. Initially, I was startled by the information as I did not see disease as part of my life lesson. I had dealt with medical problems the first forty years of my life and had overcome them. I had virtually eliminated western medicine from my regimen since most of what I had experienced through surgeries, prescriptions and invasive testing simply created more problems than it cured. I was very happy maintaining good health through my eating habits and my monthly round of practitioners for craniosacral therapy, acupuncture, Chinese herbals, deep tissue massage, and chiropractic. In fact, I had wrestled with myself about getting a pap smear, preferring alternative health care procedures which were more maintenance-oriented than reactive to problems. Typical of Western medicine, I had begun one procedure only to find I had to return for another.

I could not reconcile that I might need to devote energy to health problems given the smooth sailing promised by my guides, and the precautions I was taking in respect to my diet and supplements. Didn't they need me to be strong and capable and not distracted by health problems in order to accomplish what I came here to do? I was upset and confused by the situation which always meant I was receiving a lesson. But, what was the lesson?

Several hours passed as I continued to contemplate what I had been told. In the meantime, I went ahead and made my appointment for my kolposcopy. I did not react as I might have in the past when told that there were no appointments available for a week. In prior years, having allowed doctors to hold the ultimate authority over my body, I would have gone nuts having to wait a week to get more information about questionable results. I realized I no longer believed that physicians had the ultimate answer. Rather, I believed that many sources might have the answers to my latest health challenge, including myself!

I also wondered how serious this might be since the HMO had already waited sixty days before contacting me on a lab report that typically took fourteen days. All of a sudden it hit me! This notification was tied to a time in my life when I was being challenged to take back my power. I had always succumbed to the power of the physician, allowing them to make decisions about my body based on their non-spiritual perspective. As I pondered this new concept, I suddenly got the sense that my spirit guides were laughing and clearly heard them chide me for judging my ability to hear their communications clearly. I heard them say, "Where do you think you got the notion of coming up against an old archetype as a challenge to your power?" I silently joined in their laughter.

I could no longer live in fear of the ultimate authority figure in our society, the doctor. I needed to find answers that resonated with me. I also believed, at my gut level, that I was being tested on my faith. During the last channeling session, Devin said I had to know inside of myself what was right for me rather than seeking answers from external sources. He also said they would grant me assistance to make this journey easier. So, I asked Devin to assist in ensuring good health for me now and in the future. I knew I was making a profound, life-changing shift in regard to my attitudes about my health.

Traveling to Nibiru

I enjoyed my weekly appointments with Bobbie. This was someone who could tune in spiritually, emotionally, and etherically to me. It would be interesting to hear what my good friend and spiritual advisor received in regard to my recent health challenges. Bobbie scanned my body for any energy obstructions and told me that she saw a great deal of radioactive material in my liver. She asked me where I might have contracted it, and I recalled information Devin had provided during one of his channelings about the unusual amount of radiation on the planet of Nibiru. He said the Nibiruans were initially interested in Earth because of its gold. They had used the primate predecessor of the human to mine gold for use on Nibiru, employing a technology that used gold as a means of protecting Nibiruans from the intense radiation of their atmosphere after years of atomic warfare.

I repeated that information to Bobbie who asked my guides for further clarification. They told her that, indeed, during my sleeping hours, I was astral traveling to some heavily radiated area to conduct work with its inhabitants. I wondered if it was Nibiru? Although my physical body was asleep in my bed, my energy body was able to sojourn elsewhere, working with a colony of beings who were trying to overcome the difficulties of their planet by using my knowledge of the people, plants, animals, and minerals that existed on Earth. Obviously, I did not remember any of this upon waking since we experience Earth without all of the knowledge our High Self holds. As I ultimately discovered, we are given many clues along the way to assist us in lifting the veil if we choose. Recoding is a method for shedding light on who we are, a road back to discovering our soul identity.

As Bobbie continued to search for information, she was told I often traveled to other worlds at night. Prior to recoding, I had traveled astrally. The lighter I became through recoding, the more I was able to travel in my "etheric body" which is the blueprint for the physical body. Anything that occurs in the etheric body eventually siphons into the physical body. Apparently, this was an in-between stage in which recoding had progressed enough to allow my etheric body to travel, but it was not quite light enough to travel without picking up atmospheric influences. As the recoding progressed, I would be able to move back and forth without suffering any consequences.

I wondered if this intensified radiation had anything to do with the questionable pap smear results I had received. Bobbie and I surmised that the radiation might cause errors in reading, even radiation that was not actually present in my physical body but resided in my energy field. She called her friend, Kim, another intuitive counselor, and left a message asking her to assist us in this perplexing matter.

Summary: April Recoding Lessons

- We must formally accept our soul contract in order to recode to ensure that only Light work is conducted by those receiving power from twelve DNA strands.
- In order to recode via this particular process, we must formally request recoding from the Sirian/Pleiadian Council and receive approval. The terminology for that request is included in Appendix A: Summary of the DNA Recoding Process, Phase I, on page 154.
- Recoding enables us to be psychic, cleaning the channels that have been unused thereby increasing our clairaudience and clairvoyance.
- Recoding requires us to rely on our internal barometer, taking our power back from authority figures, the friends and family members we allow to feed off our energy, and non-incarnate beings. Since we will be receiving enhanced abilities with recoding, we need to feel confident by trusting what we receive from within.

Chapter Two

May, 1996

Originating on Nibiru

I received a call from Bobbie's friend, Kim. She had received Bobbie's message and was contacting me to share information on cleansing radiation from my body. She had gone through a similar experience while doing her own nightly interplanetary travel. Ironically, she also had received a suspect pap smear during that time and had avoided surgery by refusing to follow the physician's regime in lieu of her own remedies. She quickly advanced to a normal pap smear by using herbal supplements that cleansed her system. I listened closely as she described the natural healing mechanisms she had used. This confirmed that self-empowering methods of healing were the proper methods for me.

Kim also confirmed I was traveling to my "home planet" to conduct work with a community who was trying to rejuvenate their planet following harsh effects of radiation. I asked her if home was the Pleiades, since I knew I had lived in the Pleiades prior to coming to Earth. I also knew I had experienced seven other locations in addition to Earth and the Pleiades. Several years ago, I had been told in a channeling session that I originated on a planet ruled by a supremacy society that existed beyond our sun. I had left that planet based on disagreement with the operating philosophy. I had recently wondered if that supremacist society was Nibiru, a Pleiadian satellite the size of a large planet with a 3,600 year orbit that cycles around Earth. She told me to write down my questions and she would ask them during a cleansing she would perform remotely on my energetic body that coming Friday. We would discuss the results the following week.

The First Four of Nine Segments of Recoding

I have been trying to piece together my role on the Nibiruan team. Devin, currently the leader of the Nibiruan Council, had finally shared that he had been N's father on Nibiru. Devin was Anu, ruler of Nibiru during the time the two-stranded Earthlings were created on this planet. He had initially kept his identity from us because he thought we would find it difficult to believe. N had been Ninurshag, who was a geneticist, and had been responsible with her brother Enki for cross-breeding the Homo erectus on Earth with Pleiadian DNA to create the two-stranded version on Terra called, not too respectfully, the Lulu's. Information has already been written about Anu, Ninurshag, and Enki, particularly by Zecharia Sitchin in his series *The Earth Chronicles*. I strongly suspected I was a family member of this team but had been unable to define the relationship.

I called N and asked her to channel Anu so I could ask some questions about this. She went into a trance, and Anu emerged. I asked Anu about his choice of me as a member of the team. He said, "Dear One, you chose the project. It did not choose you." He asked if I noticed how my "heart came alive" when I recorded my progress with recoding. He was right. Not only did I crave the time to work on it during a very busy schedule, I loved every minute of it and crammed my writing into the wee nightly hours. I also asked Anu if he could define my contract and who I was vis-à-vis the Nibiruan team. He told me clues were being provided if I would heed them. He said to remember we were all connected in this plan and to know I was part of the "family." Finally, I asked if I had any medical problems to worry about. Anu said my medical problems were about releasing fear (as I had suspected). If I held onto my fear, I would create problematic situations. If I released my fear, I had no worries.

I thanked Anu for his input and thought we were concluding. Several moments passed before he spoke again, asking that I be sure to share the formal recoding request they had given me with any others who wanted it. He said someone must accept their contract prior to being recoded, a process that will ultimately render one clairaudient and clairvoyant to information from higher dimensions. The reason one must accept a contract is to ensure they will use their new power wisely rather than aligning it with self-gain. Anu also said there was no reason to grant someone recoding who had not accepted their soul contract because they would not know what to do with it. The reason for accepting a contract was to align the soul with its purpose, placing it amongst a team of assigned entities whose role was to protect that novice. Apparently, once someone is approved for recoding, they attract attention from both Light beings as well as dark forces who are attracted to enhanced power. It was necessary for the person undergoing recoding to be protected as they expanded their psychic abilities and power.

Anu continued by saying there were nine segments of recoding. According to Anu, we would not be given information on the segments until we reached them as they would have no meaning for us until then. Since N was on segment four, we were given information up to segment four. Later, we would discover that both N and I sequentially experienced the segments in order to help educate the genetic engineers since no third dimensional being had been taken through this particular recoding process. Those who subsequently experienced recoding would not undergo a linear process. They would move among the various segments, clearing and completing them in a random rather than progressive order. In other words, one might clear and complete segment three while still working on segment one.

Anu shared information on the first four segments. Segment one, Releasing Anger, represented the resolving of old issues without anger and with love, healing the physical system through this process. It would be necessary to communicate with anyone with whom we had unresolved anger. In some cases, we should actually call these people if we felt we could speak to them on a heart level. If we did not know their whereabouts or thought we might elicit more anger or defensiveness, our High Self could "speak" to the High Self of those involved. This ensured that the extreme density created by anger in the physical body would be removed prior to receiving the twelve-stranded energy. If not, pain would occur when the higher twelve-stranded frequency inter-acted with the denser frequency of the lower emotion.

The second segment, Managing Anger, involved calmly con-fronting issues while remaining in a neutral place, rather than stuffing anger over new issues, thereby creating resentment. If anger is stuffed during recoding, one will feel some discomfort. As one experiences recoding, one will discover a feeling of neutrality toward situations and greater levels of compassion. We would be tested repeatedly on our ability to remain "unstuffed," finding many situations thrown in our path which would require us to confront openly and calmly instead of with rage.

The third segment, Clairaudience, was when the telepathic channel began to open. As the lower emotions are cleared from the system, we become a lighter vehicle. It was the point where one began to speak to, and receive messages directly from one's guides. It was imperative to avoid heavy electrical environments such as video arcades or carnivals during this stage since they cause static in the telepathic channel and could result in discomfort.

The fourth segment, Clairvoyance, involved the opening of the third eye. At this point one may begin to see guides or other entities that visit or assist the physical world. It may begin with a sensing of dark forms, followed by lighter forms, moving to a full range of color images. Or, it might be experienced psychically through the third eye. It was necessary to move through any fear that arose out of this new ability due to the viewing of previously unseen entities because it would keep one from progressing through the segment.

Anu explained how important it was to avoid heavy electrical environments like video arcades or amusement parks during segment three and four. The extreme amount of electricity flowing in conflicting directions creates a clog in our circuitry. Apparently, this type of electrical flow caused so much static that the guides cannot communicate for some time after that. I asked if my frequent travel caused communication problems due to the electricity experienced during plane trips. Anu said that plane electricity flowed in a circular motion so it did not cause static compared to environments bombarded by electricity from different directions.

I wondered how I was able to channel automatic writing at my computer or receive intuitive messages despite the fact I wasn't an "open" channel. Apparently, I needed to clear the last vestiges of anger from my system, something with which I had been working for the past two years, and then I would be ready to receive continuous communication. The major emotional blockages needed to be removed prior to clairaudience or they would create physical discomfort. I remembered two people with whom I had to clear residual anger and I decided to take care of it that evening. I thanked Anu for his information and N returned.

I told N about my frustration with my lack of knowledge about my role as a member of Anu's team. I said my only clue was the name Lahaina which had been given to me in the middle of the night and probably meant I was still recalling my recent trip to Maui. N's eyes opened wide in surprise when I mentioned the name Lahaina. She said Anu had given her the names of the members of the Nibiruan Council on April 2, and Lahaina was mentioned at that time. I wondered if, perhaps, I was Lahaina in both places, a nonphysical member of the team as well as a physical representative on this planet. I tried to envision the logistics of living simultaneously in two locations, a problem bypassed by my complete lack of recall during conscious states. I wondered if I would be able to hold dual awareness of multiple states of consciousness after recoding.

It took me a long time to accept my role as Lahaina, a non-incarnate member of the Nibiruan Council, in addition to being an incarnate representative on Earth named Anne. I had difficulty viewing myself as a higher dimensional being who had a level of knowledge and power that Anne could not experience. Yet, as the days passed and Lahaina became more and more a part of my life, I grew to accept who I was. At some point, I apparently merged a part of Lahaina with Anne because Anu and Joysia spontaneously began calling me Lahaina during the channeling sessions. Ultimately, I understood my purpose was to assist the non-incarnate Nibiruan Council members in understanding how third dimensional beings would experience recoding. They needed someone with a dual perspective, someone living among them and conferring as well as someone reacting from a third dimensional point of view to recoding. I was one and the same!

Retrieving More Power

I made the decision to cancel my upcoming kolposcopy. I decided that, even if cancer cells were discovered in my body, I would not choose to have a hysterectomy or undergo the intense effects of chemotherapy. I also decided that my guides were simply testing my faith in terms of how strongly I believed I was here to help myself and others experience recoding. I decided to continue with Kim's recommended herbal cleanse for thirty days and rescheduled for a pap smear instead of a kopolscopy, assuming the results would be fine. At this point, I was still hedging my bets, wanting to have the faith to believe that all was well physically yet desiring the stamp of approval from modern medicine. However, my action was the first toward breaking a fear cycle I had allowed physicians to impose on me. I had been in and out of hospitals and doctor's offices since I was a small child, religiously following the physician's directions. I felt good knowing that I could direct the process according to my needs, rather than allowing my perceived authority figures from the past to impose their medical agenda upon my recovery.

Coming Out Metaphysically

I received a postcard from a nationally prominent metaphysical magazine regarding publication of an article I had submitted. The name of the article was *You and Your Image*, based on ideas received from my guides during an automatic writing session. The note from the publisher said they needed a short biography on me, as well as information about my source. I was unsure how to describe my guides, then thought, why not let them describe themselves? This is the information they sent me via automatic writing:

"We are a group of entities that are actively involved in recoding the starseeded interplanetary beings living on Earth. Our work will ensure a smoother transition for Earthlings when the planet shifts from the third to the fifth dimension by preparing our emissaries, the starseeds, to act as experienced facilitators for the almost five billion residents on this planet who will be given an opportunity to ascend."

Tears came to my eyes when I read the information I had received since it emphasized the profoundness of the work in which I had become involved. It also solidified my soul contract in my mind, reinforcing my role as a facilitator in physical form to assist the Sirian/Pleiadian Council with their objectives.

Lahaina "Links" for the First Time

I received a tape from Kim today, a recording of the session in which she conferred with my guides to determine where my radioactivity originated and how to clear it. She said my guides apologized for the amount of toxicity I was carrying due to the radiation I was picking up during my travels to my "home" planet. They had apparently underestimated the vulnerability Earthlings had to radiation, having boosted their own immune systems eons ago due to the greater amount of radiation in their atmosphere as a result of warfare. In fact, Kim said the entire recoding process had been slowed based on this finding. Apparently, this was the type of information the Nibiruan Council needed from my third dimensional body since it was a reaction to recoding. In fact, Kim was told that the Nibiruan Council was working to alter this response since it might occur for other recoders, resetting "dimensional coding" to allow third dimensional beings to travel to "interplanetary stations" without picking up radiation.

Next, Kim spoke of my origins. She said my soul was conceived on Nibiru from a Nibiruan female and a Sirian male. They had come together for the express purpose of creating my energy field to, in Kim's words, "create a new form to help save the beings from both my planet of origin and Earth." This was shocking information for me since I hardly viewed myself as a prophet and certainly had no desire to be a savior. However, the information seemed to fit with the higher aspect of my-self—Lahaina.

Kim also said I was called a "mutable" because my energy field was extremely adaptable to differing planetary systems. I was also called a "gatherer" which meant that, even though I had originated on Nibiru, I had resided in eight other locations due to my "mutability." To name a few, I knew I had lived with the Jolanderanz who look like the underwater beings in the movie *The Abyss*. I had lived with the Pleiadians and even relocated to live with those almond eye abductors called the Grays as part of a research experiment that was Pleiadian-based. I also knew I was a member of a Sirian group who had resided on Jupiter. Sirians are sixth dimensional beings who live in etheric form. This combination of "mutable" and "gatherer" lineage created an ele-ment of fragility in my energy field since my matrix was constructed loosely to enable interplanetary traveling and communication.

It occurred to me that perhaps my soul father was of a different form than the Earthling form to which I was accustomed. Perhaps he was similar to this Raphael energy I had carried in my womb for eons to release upon Earth to help people at this time. I asked my guides to find out if this was the case and was told that the energy was similar, but the form was not. As I learned, Earthlings originated from Hominoid form but there are many forms of Hominoids, some human in appearance and some not so human. It is like comparing a whale in the ocean to a woman. Both are mammals, yet both have different forms. My soul father's form, albeit Hominoid, might look different. As a gatherer, I had enjoyed many aspects of physical and nonphysical experience. It was time for me to understand our souls existed beyond human form, and I needed to love myself in any form, regardless of how unpalatable it felt based on Earthly standards. This notion brought new meaning to the idea of acceptance.

Kim also provided a piece of information that explained the simultaneous multidimensionality of Lahaina. She said based on the flexibility of being a mutable, I had the ability to break off a part of my energy and coexist in two or more dimensions simultaneously. It would be easy for me to live as Anne on this planet while acting as Lahaina in another dimension. In fact, it was possible that Lahaina was acting as the orchestrater and overseer of Anne. It was taking time for me to process, understand, and accept this idea.

Next, Kim informed me that my people saw me as a "hero." This saddened me as I did not feel this label fit. I wondered what I had done to deserve such standing, feeling weak compared to Lahaina's power. I have been without conscious knowledge of my abilities for so long, I had no memory of them. When I initially came to Earth on this "assignment" 120,000 years ago, I am sure I envisioned a much shorter time line. I must have given up hope along the way, thinking I would never recover my abilities. Now that it was so close, I felt consumed by the need to recode and gain broader consciousness.

Kim then gave some additional information on why Earthlings needed to accept their soul contracts before they could recode. A soul contract is a sacred agreement made between a being and their higher galactic council, in our case the Sirian/Pleiadian Council, prior to incarnating. The process of recoding works in concert with realizing one's soul contract. If someone knew their contract before undergoing recoding, it would seem meaningless. They would not have the desire to manifest it until their vibration aligned with the same vibration that originally created the contract. In essence, we must somewhat blindly agree to something we do not presently have the capacity to understand, trusting that our soul knew what it was doing. Having agreed to something beyond our comprehension, we are then awarded the return of our abilities to realize the intended outcomes of the contract.

Finally, Kim explained why I was having difficulty channeling, a step beyond my present telepathic capability of automatic writing. She said since the matrix of a mutable was loosely constructed for purposes of "shape shifting," it would be very uncomfortable to channel because I would have to hold my energy at a dense level for a long time. Channeling relies heavily on the fifth chakra (throat), allowing the entity to enter one's body and use fifth chakra energy to communicate. That process works contrary to the makeup of a mutable. Kim said I needed to learn a process called "linking" which would allow my guides to speak to me with great clarity. In fact, she said my guides felt I was ready for linking and had already selected an angel named Nephras to communicate with me.

Linking would use my telepathic setup rather than my fifth chakra energy. Apparently, my telepathy was well-developed as a mutable and was constructed to have a broad range of frequencies. This meant that I could interpret a wide variety of languages. Kim instructed me to link through my Alpha chakra, an eighth chakra that was situated above our seventh crown chakra, about a foot and a half above the head. She said the only reason I could not link today was because my Omega chakra (located eight inches below the root chakra in the mid-thigh region) and Alpha chakra were out of balance. I needed to tune them by holding one hand over the Alpha region and the other over the Omega region and circulate energy back and forth to bring them into balance. After I balanced, I would be able to receive energy through a universal telepathic channel that uses the eighth chakra. Kim said the information would flow slowly at first, and I might want to use a "yes" and "no" question format. I was so excited by this information that I could hardly wait to get started.

That evening, my friend Pat came for a visit. I had asked her to assist me in linking. We lit candles and asked our guides to come forward and assist in harmonizing the Alpha and Omega chakras. I placed my right hand over my Omega chakra and my left hand over my Alpha chakra and began to envision a clockwise circle of energy moving between the two areas. Pat envisioned the same.

After several moments, I needed to rest my arms. I had a pressing sensation behind my ears and a tingling in the bottom of my feet. In fact, the tingling has continued to persist and gives me the sense that I am continually "turned on." I asked Nephras to present herself. I could hear her clearly. Nephras would adeptly drop an idea into my head which I would then translate. The concepts were accompanied by supporting images.

We began with easy questions. Pat asked what she should do about her dying dog, Abby, a companion she had enjoyed for eighteen years. She said she had attended a dog show in Wichita the prior month and felt a strong bond to a Cocker Spaniel named Cici. The owner was keeping Cici until Pat decided if she wanted her, feeling that Cici and Pat were the perfect match. I immediately saw a type of dog communication system that allowed animals to speak with each other telepathically. Apparently, Abby had placed a dog version of a "classified," requesting a new body for her aging vehicle. I could see animals were different from us because their souls reside in a collective pool. Later, this was confirmed for me by A. E. Powell's work, a noted early twentieth century metaphysician. Animals can move in and out of their collective pool at will, not needing to wait until death to vacate their body. Cici had responded to Abby's call, offering her body as a vehicle.

I was laughing as I repeated this information, thinking it seemed silly, but Pat was pleased. She said she had sensed she needed to bring Cici home but had resisted. I told her Abby was waiting for Cici's arrival to ensure she had the new body form before she left her aging body and would leave (die) before long. Cici's soul would return to the collective pool. Pat decided to drive to Wichita and get Cici. She then asked if she would need to put Abby to sleep or would she die of natural causes. I hesitated answering because I knew Pat preferred her to die in her sleep, but I kept seeing one of those metal veterinarian cages in which they housed animals. I simply told her what I saw.

Pat asked me several more personal questions, warning me not to edit what was coming into my head since judgment would shut off the flow. We spent about an hour linking, then Pat asked if I would assist her with her recoding request. I closed my eyes, requested Nephras to help us, and asked Pat if she agreed to accept her soul contract. She said "yes" without reservation, and I heard Nephras say that her contract was to "sweetly heal" others, although not to underestimate the difficulties she might undergo when facing the dark forces. These forces were our illusion of darkness which attempt to hold us in its grasp through our emotions of fear and guilt. Pat read the formal recoding request. Then I saw this beautiful display of lights and heard clapping and cheering. She had been accepted. Pat thanked me for helping her to come to peace about her dog, Abby, and then she went home.

Later, Pat went to Wichita to get Cici and await Abby's transition. She watched as Cici and Abby exchanged energy back and forth in preparation for Abby's transition. Literally, they would swap personalities which would have completely perplexed Pat had she not been forewarned of their plan. Pat's other dog, Alex, became very uncomfortable with the process, liking Cici when she held Abby's energy, but causing havoc in the house when Cici reverted back to her own energy. Eventually, Cici decided not to swap with Abby because she fell in love with Pat and did not want to leave. She had also developed a similar eye disorder that Abby possessed, so Pat decided to take the dog back to Cici's breeder and graciously accept Abby's death.

It would take several months for me to gain confidence in my new psychic ability, but eventually I felt like I was receiving information that was not only accurate but extremely beneficial and healing to the recipients. In fact, my psychic acumen continued to grow until I expanded beyond linking into full channeling, working within the confines of my body type yet able to receive a full spectrum of information. As my confidence grew based on the positive results I saw people receiving, I established a business called InterLink which was devoted to assisting people remove blockages that stood in the way of expanding their consciousness, a role that aligned perfectly with my soul contract. Finally, I was consciously and effortlessly channeling, maybe not in the way I thought I would like a trance channeler, but in a way that was in alignment with my unique energy.

The Concept of Copies

I called N today. She was having difficulty because her astral body was separating from her physical body, unfortunately while awake instead of during sleep when it would be less startling. Apparently, her guides needed to work on her by taking her astral body for several days to do the recoding work, leaving a "copy" in her place.

As I eventually came to experience my own "copy," something everyone would undergo in recoding, I grew to learn we needed to take responsibility for training our copies to our standards. Copies were necessary because our astral bodies needed to undergo successive days of treatment by our genetic engineer as we became more involved in the segments. They simply could not complete all of their work during several hours of sleep each evening. However, copies tended to indulge rather than act disciplined. For example, my copy did not like to exercise, a routine I maintained since it reduced stress and kept my weight down.

I reflected on the incredible overload a linear thinker like an Earthling would encounter if total recall of our astral body experiences was possible, especially since the separation occurred during waking hours for recoding. It would be like watching two movies at the same time, hearing and seeing two different story lines while living in a single physical body. I decided I was not prepared for this experience, yet it was a taste of the fifth dimensional state I would enter in the future.

Altering Frequencies to Recode

I was very sad. My relationship with Jerry was ending, and it was with great difficulty that I released our union. We had been together many times, and we had planned to be enjoined for the transition to the fifth dimension. However, it was not working. The more I moved into my new way of being and new way of thinking, the less we seemed to have in common. This was hard to accept due to the love I felt for him. However, I recognized that I was in a holding pattern trying to preserve a dying relationship. I felt tremendous conflict. If I moved forward spiritually, I harmed the relationship, and if I moved forward in the relationship, I lost my spiritual gains.

This decision wasn't easy. I often wanted to forget what I had recently learned to go back to my "old" ways, even if they were uncomfortable. But I found the disharmony and the lack of truth for my path were taking their toll. I think I knew subconsciously when I accepted my soul contract that this was an inevitable outcome. This is why I asked Laramus if Spirit could make transitions easy on me rather than "forcing" me to change. I remembered him saying that the decisions would evolve naturally, and I would only change when I was ready. After all, we live in a free will zone and Spirit cannot subject us to anything unless we desire it. He omitted the part about the distress we experience from a life change decision, no matter how "ready" we are to move to new territory.

On top of the relationship problems, I had been terribly ill over the weekend, ending up in the hospital emergency room due to a virus located in my female organs that caused a one hundred and four degree fever. Apparently the stress caused from the impending loss of my relationship, and the reconciliation process with Spirit took its toll on me. Since my energy had changed, it was uncomfortable for me to live in conflict with the intent of my soul contract. I felt no other choice but to follow Spirit. I allowed the illness to rid my body of an immense amount of toxic radiation and stuffed emotions. It was now up to me to expedite healing through natural mechanisms.

I decided to link in order to better understand the reason for the level of physical pain I had experienced. I was not yet ready to receive information on my relationship since I was sure I would not like what I heard. My guides gave me the following information:

"Dear one, you are going through a difficult time that you didn't expect. You will survive quite nicely, although it doesn't seem like it. We have told you, yet you still don't quite believe it, that this purging of your system is the lesser of evils you would have had to endure. Your earlier years were filled with high levels of negativity which no longer mesh with your energy. You had to rid it from your system. Your prior emotional cleansing work has helped immensely because it cleared you in the area which held the most anger—your reproductive system. You have lived with this anger, not only from you but also from your mother who has felt limited as a female in a male dominated world. She subconsciously knew she was carrying a daughter in her first trimester and transferred that anger to you, thinking you would not realize your full potential as she had been unable to do in her world of the 1930's and 1940's. You have sat with this energy in your own female organs, supplemented year after year with the gender imbalance of your society. Now, it is gone. You are truly cleansed and will have no more problems. Your residual discomfort will eventually stop. It is part of the balancing, and it is also a reaction due to a weakness inherent in this part of your body. We knew this was coming and apologize, but it was necessary. You would have manifested illness if you had not cleansed due to the discord of your new energy with the old. This is why the first step toward recoding is the release of anger since it has the potential to cause discomfort due to the discord."

I was fascinated by this response. First of all, I knew my mother would truly be dismayed to learn that she had transferred any negative feelings to a baby she had waited four years to conceive. And yet, it was integral to her and the society which had thwarted her development. I began to fathom the extent of the difference in energy I would be experiencing as part of recoding. I was moving into an energy that was more ethereal than physical, that resided in love and joy as opposed to fear and anger. I could see how this energy would be discordant with any old, negative energy residing in my system. That continual tug on the old energy from the new energy would create a pressure system in the physical body that would have to release itself. This was why releasing anger was necessary to moving forward. I considered my bout with pain a true success and the marking of a major hurdle overcome.

Using Holographic Repatterning™ with Recoding

I was honored to be able to assist another person, Julia, to begin recoding. Julia and I met at a Holographic Repatterning class, a technique that takes negative energy patterns and transforms them into positive patterns via energy healing modalities. It works with the energetic blueprint that surrounds our bodies, reprogramming information that entered our energetic bodies during past trauma. Before I assisted Julia with her recoding request, she repatterned me in order for her to gain more experience in our new healing techniques.

Julia went through a six step process that was designed to identify the negative frequency by fact-finding the core issue. Kinesiology was used to determine the accuracy of the fact-finding.

Kinesiology is a method of finding the body's truth by imposing pressure on muscles and checking the strength of the resistance (often referred to as muscle testing). Julia and I were led to a past life issue for me that dealt with a lack of forgiveness.

I immediately knew what it was, a lifetime in Sedona when it existed as a crystal city. I always saw this crystal city hovering above the red rocks while vacationing there many times. I knew from past psychic readings that I had been put in charge of the sacred codes for that ancient society. My selection had been based on the society's trust that I would make decisions framed by the principles of love. I had faced a difficult time, because I was put in the position of protecting the society's destiny by keeping the codes from invading Reptoids. I had learned about the upcoming invasion based on my mutable composition that enable me to communicate with others. This made me privy to information no one else could access. In order to protect the codes, I hid them somewhere under Sedona and "blew up" the area, destroying the society that I had loved in trying to prevent an even greater disaster. Apparently, I still know where these codes are and I am to discover them in this lifetime when I gain consciousness regarding their meaning and whereabouts. In the meantime, I had experienced great pain over many lifetimes due to the decision to destroy a society I loved.

The last time I visited Sedona, I sat overlooking Boynton Canyon, crying and repeatedly saying "I'm sorry." At the time, I was unsure why I felt sorry. Yet, I was wracked by pain. Later I learned through my guides about the decision I had made that destroyed something so dear to me. When I told Julia about the Sedona experience, I started sobbing.

I recovered from my emotional outburst and proceeded with the repatterning. I had to diffuse the lack of forgiveness by first experiencing it and then breathing it out of my root chakra. My healing modality consisted of placing various tuning forks on my nine chakras, the traditional seven, plus the Alpha and Omega points. I could see why I had difficulty resonating with channeling. When I had been privy to other-worldly information in another lifetime, it led me to make the decision to destroy that society. I carried a fear of accessing ethereal information since it might drive me to equally painful decisions. I was finally cleared from that guilt and pain and was ready to link for Julia and others.

Then, it was Julia's turn. Julia's reading flowed easily and apparently brought her great comfort as she had several life change decisions to make as well. The beauty of her soul contract was also brought forth to her in a memorable way and fit nicely with her talents and interests. She kept exclaiming, "how perfect!" as I answered her questions. I was pleased with this response since we had never spoken personally, and I had little point of context for interpretation. I kept gaining experiences that built my confidence about my telepathic skills.

I had wondered why I had taken the Holographic Repatterning course at the time and was now beginning to understand. When we recode, there is much baggage from past events that stands in the way of the new, lighter energy that is coming into us. The repatterning process can clear emotional scars from past lives, prenatal experiences, and our current life within minutes, eliminating the need for years of therapy to erase those patterns. As we approach the final days before the shift, we do not have the luxury of time to experience lengthy therapeutic treatments. We also need to ensure that we have eliminated negative energy from our system that goes beyond the sometimes superficial healing we undergo when dealing on a mental level. Holographic Repatterning had been offered to me as the ideal solution.

Releasing Toxins in Order to Recode

I met with N today in order to speak with Anu and obtain confirmation of what I had experienced to date with recoding. I was in segment three, Clairaudience. The session began with Anu stating that there was "another" who was jumping with impatience to speak with me but that the "other" would have to wait until Anu had finished. I laughed and asked if it was Laramus. Anu said, "Yes, it is your old friend Laramus who awaits with great impatience." He asked if I would like to speak with Laramus first and I said, "no," I would hear what Anu had to say prior to communicating with Laramus. I figured that I should apply the first-come, first-serve policy to nonphysical beings since it was the way I liked to be treated in physical form.

Anu began by telling me he loved me and held me in the highest regard. He said he was proud of the way I had decided to move through the recoding process without fear, leaving doubts behind me as I held onto trust and faith through the rocky emotional and physical events of the past few months. He recognized the difficult emotional lessons I had been learning during recent weeks and told me to continue to remain true to my path. He said I was surrounded by assisting guides at all times, and I was very well-protected.

I asked Anu about my recent viral infection and its purpose. He confirmed that the infection was the necessary elimination of toxins and negative energy in my system, assuring me it would not be repeated. He said recoding required the use of an entirely different energy wave. And, if it met with discordant energy, it could create sickness if someone does not clear sufficient negative energy prior to recoding. People who move forward prematurely may have some nausea, fever, headaches, and low energy. Also, anger had to be released prior to receiving the twelve-stranded energy, or there could be discomfort as the lighter energy met with the denser energy of anger in the physical body.

I asked Anu to give me a sense of how I had performed during my linking sessions when I had assisted my friends in asking for recoding. He said I had done very well and just continue what I was doing. I asked if I should do some type of advertising to let others become aware of my assistance. He said I should await the requests from others through word-of-mouth at this time. He then referenced the nine segments of recoding, stating that some of those who came to me for recoding information would not move through the entire nine segments. Some, mostly starseeds, would probably proceed through nine segments. Many of those whose souls had always incarnated in the Earthly realm would possibly only make it through the initial few. However, Anu said that each would attain the segments needed for their evolution at this time. In other words, the segments we attained would either match or exceed our soul contract.

I had suspected for some time that I had originally been part of Anu's Nibiruan family. I asked Anu if I held any relation to him. He confirmed that my soul creation was tied to members of his family, saying that I needed to speak to a genetic expert about my lineage if I wanted more information. I asked if that would be possible during this particular session and he said, "yes," after I spoke with Laramus.

Anu exited, and Laramus emerged. He said he was most anxious to speak with me since he knew my personal relationship was rocky, and he requested that I "simmer" for a period of time. I guess he knew I had decided to share my concerns with my partner, Jerry, concluding through much grief and pain that it would be best for us to pursue separate paths. He said I had the right to exercise free will, and he respected my decision to communicate my feelings. He requested I wait a short time before acting since Jerry's guides were actively working on a certain "barrier" that he needed to cross. If he chose to cross the barrier, we could complete our work together. If he did not, we would need to part. I told him I would wait a brief interval based on his input, but that I desired to resolve the situation shortly since it potentially involved changing my residence. He thanked me, reminding me that part of the lesson I was currently learning involved patience and tolerance.

I asked Laramus when I would be able to communicate with him directly like I could with Nephras while linking. He said I needed to proceed two more segments in recoding prior to being able to communicate. I asked him to share those segments, and he said he could not provide that information in advance of experiencing them. I always tried to wheedle more information out of these beings, but they liked being mysterious. Later, I would understand that they simply did not want to rob me of my own experience by predetermining or predicting my future. I also grew to appreciate their stance since we tend to create within the limitations of our perception, and I might have altered my experience, perhaps restricting it, had it been defined.

Next, I asked Laramus to shed light on the relationship between the Nibiruans and the Pleiadians, now knowing I had lived in both colonies but originated on Nibiru. He said that Nibiru was an artificial planet made by the Pleiadians for use as a spaceship. He compared Nibiruans and Pleiadians to Americans and Texans, explaining that they both were the same. He also said the Pleiadians had frequently cross-bred, but the original, pure Pleiadian strain was Lyran and produced eight to ten feet tall beings with blond hair and blue eyes. Out of curiosity, I asked him how many lifetimes I had incarnated since my soul's inception into physical form. He said I had lived 836 lifetimes, 362 on Earth, and the remainder on eight different planets and a variety of lesser known stars as an etheric fifth and sixth dimensional being. Knowing the Nibiruan year was equal to 3,600 of our Earth years, I assumed that the disproportionately high number of lives on Earth was probably due to our short life expectancy.

Laramus shared his love and left, allowing Joysia to come forward. I had never spoken to the chief genetic engineer, Joysia, but I had questions regarding recoding and it had been determined that I should speak to a resident expert. I asked if I should be introducing my son, Drew, who was eleven, to recoding. Joysia said not to worry about Drew since the children would be protected during the Earth's shift regardless of their progress through recoding. He said Drew would begin recoding at the appropriate time.

I knew Drew would need to eventually recode. He was an exceptionally spiritually gifted child as I had learned thorough channeled information I had received over the years. He was part of a soul group of twenty-three beings, some incarnate and some nonphysical, who had a soul destiny contract of some kind with the planet. He also had thirty guides and teachers observing him, placed there for his higher good according to the input I was receiving. I knew he would be involved in work that would heal others. It was just a matter of time before Drew would be attracted to recoding.

Next, Dyjouniaa, the Keeper of the Records, emerged to answer some of my questions regarding my lineage. I asked him to tell me about the inception of my soul. He said my mother was one of Anu's daughters, and my father was a "great one," a leader among the Nibiruan's. He said my father currently was an eleventh dimensional entity who would eventually like to speak to me. He acts as a "consultant" to the Nibiruan council, the group of entities lead by Anu. Dyjouniaa said my father was part Sirian and part Nibiruan and my mother was part Sirian and part Pleiadian, intentionally mating to produce my body type which would carry multiple codes. Dyjouniaa said matings were done with precision and care on Nibiru since DNA codes were very precious. Each individual was inseminated with specific combinations of codes in order to ensure the development of specific talents. I had been created with much forethought because my codes would allow me to communicate with many beings from different planets. In fact, he said I was frequently taken on space missions since I was the communicator. He called me a special child who could decipher the codes like a computer. I wondered when in the recoding process I would be able to regain this incredible talent since Dyjouniaa said I retained this ability, capable of tuning into different people on different frequencies just like a TV has multiple channels. I asked Dyjouniaa if my father had looked like a human since I felt he carried the same energy as the Raphael being I had brought to this planet as a seedling. He said my father was human but had slightly different appendages than we know. My mind was bulging with all of this new information, and I decided it was time to end the session.

Summary: May Recoding Lessons

*   The Nibiruan Council is a key player in the DNA recoding process based on past karmic ties due to their involvement in creating the two-stranded vehicle. The two Nibiruans involved were Ninurshag and Enki. Both Ninurshag and Enki are children of Anu who heads the Nibiruan Council. [Note: This earlier two-stranded version was very limited in its functions, i.e., it could not propagate or mentally process information in a cognitive fashion, and was eventually replaced by Adamis which is the predecessor of modern man. Both N and Anne agreed to incarnate on Earth. This

was in addition to their oversouls maintaining a presence on
the Nibiruan Council to assist the Council in understanding
the third dimensional thought process as well as to physically
undergo recoding as a trial run.]

- Joysia is the chief genetic engineer that acts as a consultant
to the Nibiruan Council. He and his team are responsible for
actually reconnecting the twelve strands of DNA to our astral
body.

- Recoding is a process of trust. We must let go of our fears
of the unknown, because fear causes discomfort when
moving to a higher frequency. Recoders have nothing to fear
and everything to gain from the process.

- The process of recoding works in concert with realizing one's
soul contract. We may need to begin recoding without
knowledge of our soul contract since we sometimes cannot
align with the vibration that created that contract until we
have experienced a lightening of our energy through the
recoding process.

- Not everyone will complete recoding since some will not
have the commitment to move through the ups and downs
associated with the process. Recoding requires a total house
cleaning of old issues and complete faith on the recoder's
part when things potentially appear worse before they get
better.

- Those who are recoding are surrounded by energetic
protection in order to ensure they successfully complete
recoding should they so desire.

- There are nine segments to recoding. Although N and Anne
moved sequentially through the segments to support the
learning process for the genetic engineers, few will
experience the segments sequentially. Ultimately, all nine
segments must be cleared and completed for one to
energetically attain their twelve strands.

- The first four segments of recoding are Releasing Anger, Managing Anger, Clairaudience, and Clairvoyance as follows:

  Releasing Anger:
  Anger must be released in order to eliminate negative energy prior to fusing the DNA strands. This minimizes the pain as the higher energy level of recoding enters the denser energy level of the physical body. One should make peace with those who have caused anger in the past. This release can occur in person or at a soul level.

  Managing Anger:
  If anger is stuffed during recoding, one will experience discomfort. Segment two ensures that anger does not rebuild in the system after clearing it in segment one, which would re-create a level of density that would cause discomfort. As one experiences recoding, one will discover a feeling of neutrality toward situations and greater amounts of compassion. However, one should not expect to completely remove anger since anger is an aspect of the Earth experience.

  Clairaudience:
  The telepathic channel begins to open in segment three. As the lower emotions are cleared from the system, we become a lighter vehicle. This is the point at which we can speak directly to our spirit guides rather than getting the information via the dream state, intuition, or another individual who has psychic abilities.

  Clairvoyance:
  Segment four opens the third eye so we will be able to see guides and other entities that are visiting the physical world. This is a gradual process since the third eye has atrophied through lack of use in the third dimension. Clairvoyance may begin with the sensing of dark forms

followed by lighter forms and, finally, a full range of psychic color vision. Or, it may occur entirely in the third eye via psychic vision.

Chapter Three

June, 1996

The Ups and Downs of Recoding

I called Pat to forewarn her of what I was experiencing due to lingering negative energy in my body while recoding. I felt responsible since I had informed her about recoding. She laughed when I called, claiming that the last several weeks of her life had been particularly interesting since many of her old, unresolved issues had resurfaced. At first, she thought she was experiencing a backslide into poor habits. However, after consulting her guides, she determined she was undergoing a final purging process. I confirmed her conclusion, explaining the necessity of releasing conflicting energies in order to make room for the new higher energies. We sympathized over how uncomfortable it felt to repeat patterns we thought had been resolved. On the other hand, knowing we were going through a final release of negative energy patterns, we blessed our progress and resolved to endure.

Leaving Jerry

Jerry and I decided to separate. Well, so much for "simmering" per Laramus' instructions. This was very painful to me, yet I knew I must honor my growth. Jerry was looking for a place to live while I continued to work on releasing anger and grief so it would not slow my recoding progress. I was angry at Jerry for rejecting a spiritual way of thinking that would have kept us together. And, I grieved losing my intended life mate, even though we were moving in conflicting spiritual directions.

In an effort to understand my strong desire to choose recoding over my relationship, I queried my guides. Based on what my guides shared, it seemed it required blind faith to reach my objective since I had no point of reference toward which I was navigating. I must trust my intuition and my soul's knowing that the ultimate result will be for my highest good. Here is what my guides shared with me:

"We feel your emotions. They are very intense and heavy. You must take care not to make yourself ill with this feeling. Try to lighten it with laughter when possible. Sometimes lessons take you to difficult places. We are holding as much of your energy as we can to lighten the load. You are on the threshold of all that you have longed for although we know it doesn't feel like it. You will experience such profound joy that you will look back on this juncture and feel like it was utter nonsense. This is not a long term path to the threshold. You are standing on it right now. Don't turn back by hanging onto old patterns. Move forward and those who wish to come with you, will indeed. We are asking you to be very brave by moving forward with blind trust. But you have no choice, do you? There is nothing for you where you are. You cannot find anything but the emptiness you have always found because the third dimensional Earth experience promises nothing but emptiness in its current mode. Jerry is not ready to go in the same direction at this time. He has not chosen to do this. We have pushed on him so he would get serious because he was to be your life mate for this experience, but he prefers to stay where he is at this time. You are not to blame for his decision. We tried to assist him because he had a soul contract to accompany you. Now, you see how all souls do not move ahead just because they have agreed to a contract. We suggest that you understand this deeply because there will be those who will not choose to move forward in recoding. This is not your failure. It is no one's failure. It is simply choice."

In addition to the above, they told me they did not wish me to go through recoding alone. They felt I needed male energy to balance me during this process and to support me in my future work. They said the love frequency exponentially expanded the new energy we carry, magnifying it well beyond a single person's capacity by creating a dynamic interaction at the higher energies. A love frequency would intensify the effect I had on others as I evolved through recoding, improving their chance of succeeding by raising the overall vibration. This would be true for everyone experiencing recoding, not just me.

My guides said they knew I was grieving the loss of a loved one. However, when I was ready, they would send me another mate. This time, my mate would be a recent walk-in. A walk-in is an entity who, through a contract with another soul, exchanges places. This is not a demonic possession since the contract has been negotiated between the two souls. In all cases, the work of the original soul has been completed on this plane and it is time for the soul to return to spirit. Essentially, a walk-in uses the physical body of a soul who has died but bequeathed their body to another. The walk-in lives with the experiences and memories that are held in the original soul's body while bringing their own memory base.

My guides were sending me a walk-in because my mate would need to be someone who could easily access his starseed origins based on the nearness of his sixth dimensional state to his recent third dimensional state. Essentially, there was no time to train someone. This walk-in would rely on me to prod his memory, immediately moving into his own recoding evolution being in sync with my own. They also wanted me to be with someone who perfectly matched my energy field so there would be no conflict. We could focus on our mission rather than working through karma. I was somewhat cheered by this information, feeling perhaps I had not failed a relationship as much as I simply needed to keep moving. I told my guides I would let them know when my heart was ready to receive a new life partner.

Lahaina's Unique Grid Work

Julia called to tell me she was separating from her husband. This news scarcely surprised me based on my own experience. It appears that recoders will have more success intimately interacting with other recoders due to the match of similar energy fields. I offered to share my house with Julia to emotionally support each other during our transition. She decided to move in when Jerry moved out the following week.

Julia came over to find out how she would arrange her furniture, and I asked her to repattern me. I had used kinesiology to see if I resonated with a "romantic" love energy, wanting to ensure success for my future relationship. I resonated with a love energy, but did not resonate with "romance" since I had always primarily experienced life from a mental level. Julia agreed to assist, enjoying opportunities to practice repatterning. She decided to repattern me on all interdimensional levels rather than my third dimensional existence since I had told her I experienced Lahaina's existence in another dimension. I agreed, not realizing how complete a healing I would undergo during our intense two-hour session.

Julia and I uncovered the impact of a past life on my love energy. It was a lifetime where I had traveled with a team of Pleiadian scientists to live among the Grays (those blue-gray entities with the almond shaped eyes who have abducted Earthlings for experimentation). I lived among the Grays for several lifetimes, as had Jerry, but had chosen to leave when their philosophies no longer aligned with mine. I was living in much pain in that lifetime, apparently resulting from DNA experiments to which I had personally subjected myself.

Who knows why a soul agrees to uncomfortable experiences, but I assume we possess a sound rationale for enduring them. However, I must have underestimated the pain I would endure. During a past life regression I had done several years ago with Bobbie, I reexperienced my rapid exit from the Grays. Jerry, was attempting to help me leave, but based on the pain I was experiencing, I grew tired of waiting for him to determine the best exit strategy. In my intense desire to leave the planet, I had "goofed" by taking matters into my own hands, blowing my grid structure apart. Our grid structure houses our "hard drive" which essentially runs our programs and holds our past, present, and future records. By damaging my grid structure, I had altered my operating program. I saw myself explode into a mass of red lava during the regression. I confirmed through later channeling sessions with Bobbie that I returned to the Pleiades. I spent seven lifetimes there before coming to Earth 120,000 years ago to experience my third dimensional form.

While Julia was attempting to uncover the information required to complete the repatterning, she was told that another being was involved with my exit from the Grays. I heard my guides say it was my future life partner. It seems that Jerry and my future mate were busy debating my best exit strategy, wanting to return me intact to the Pleiades. However, they could not come to agreement, and out of frustration, I left on my own initiative. This is probably why Jerry made a soul contract to assist me during my current transition since he had not been able to help me in that past life. How appropriate that he was to be replaced with the other player who had also desired to assist me.

Through a combination of the repatterning method and our combined telepathic skills, we decided I had severely damaged my grid structure in that explosion. Essentially, I was not aligned with my true self. In fact, I was still sporting the same grid injuries after all these years. In addition to the grid damage, I was carrying several negative emotional patterns in my current life that were interfering with my ability to resonate with romantic love. First of all, I had isolated myself with intense self-sufficiency, not allowing others to help me. This reflected my feeling of really needing someone to help after being blown to pieces in a hostile reality. Yet, no one was there. Secondly, I carried an underlying fear of the future since I had lost my unique communication faculties, unable to relate as I had in the past. Julia said she could see my grid and that it was different than any she had seen before, a gold net interspersed with jewels. It was interesting, and confirming, that Bobbie had seen the same gold net and jewels several months before.

Julia told me to lie down. She held her hand over my heart chakra to hold the energy as my guides repaired my grid. During this time, I envisioned the explosion I had witnessed during my earlier regression with Bobbie. I felt my vast neediness when no one was there, and my ensuing fear when considering facing the future being handicapped. While laying there, I realized the implications of damaging my grid. Apparently, I used my grid to communicate with the other planets. It linked me to thousands of different languages. The jewels were the minerals that resided on the planets and stars with which I was communicating, each one representing the universe from which it came. For example, quartz crystal was the mineral for Earth since it is a primary component of the planet. These minerals are like crystals in radios, transmitting the electromagnetic energy and the translation codes for the matching locations. No wonder I had developed a fear of the future. I knew I had lost my talent for communicating which had been carefully created during my soul inception on Nibiru many years before. I do not know what prevented my guides from mending the grid prior to this time. I only knew that I needed to be consciously aware of the infraction and request the repair.

Julia had been using kinesiology to obtain answers to some of her own questions during my grid repair. She was trying to find out if anyone had a similar grid to mine. She said my guides wanted me to know I was "special and unique," since none on Earth had this type of grid structure. However, there were others like me in nonphysical form. I asked Julia if my future partner would align with my energy, thinking that Jerry may not have matched. Julia checked and said our grid structure would totally align, and we would recognize each other immediately. I wondered what it would feel like to be so similar to another since I had lived the majority of this life feeling alone and different, never understanding why I could not connect with others. It had taken many years for me to learn I was a starseed. Once I understood I was in the small percentage of beings who had not originated here, it was easier to handle feeling different. But, I had never overcome this feeling in my intimate relationships. Perhaps, that time of isolation would finally end for me when I found my life partner.

Julia completed the healing by having me do some breathing as I cried and cried, releasing the pent-up emotions carried for so many lifetimes. When we were done, I felt like a different person. I held no grief nor anger toward Jerry. This was an odd feeling since I had been consumed by these emotions. But, they were gone. I felt optimistic and positive. I felt one thousand pounds lighter, eagerly anticipating the future. I was amazed at the transition from depressed and miserable to joyous and energetic, bounding with life. I hugged Julia and thanked her for the immediate relief.

The Pupil becomes the Facilitator

Bobbie called and asked to use my new, ever-strengthening telepathic abilities to do a linking session for her. I was amazed since she had been my teacher for five years and was asking me to examine something for her. She had been dealing with a highly emotional issue and needed an unbiased and objective interpretation. I told her I would gladly assist.

We explored Bobbie's problem which, to no surprise, involved her need to separate from a man she had been dating. Our genetic engineers were doing a major housecleaning on us as we recoded, which we were not necessarily enjoying. During our linking session, I received some answers regarding her resistance to recoding. Instead of viewing recoding as a chance to improve the Earth by increasing her abilities, Bobbie saw it as a potential severing of her Earth ties by aligning her more with Spirit than with physical properties.

I telepathically viewed her past lifetime as an Earth Goddess and understood her hesitancy. She was at one with the Earth during this period of time, a true Earth spirit. I watched as she danced across fields, living a life of joy. I also saw how she could not regain that alignment with Earth in its current fear-based condition. Her past life as an Earth Goddess had been free of fear. The only way she could recapture that feeling was to improve the energy of the Earth by raising it to a higher frequency and removing fear, a process available to her through recoding. Ironically, she would become more grounded by raising her frequency instead of being less grounded on the Earth plane. This would be an important concept for all recoders to understand. We were not recoding to expedite our exit from the Earth. We were recoding to become more at one with Earth energy.

Bobbie read the recoding request, and her query soared energetically to the top of a mountain where a group of elders in long robes stood by an altar. They seemed amused when they determined the originator of the request, recognizing Bobbie immediately because she worked with them daily as an incarnate Earth member of their team. They sent approval of her request down the mountain by way of a white dove. Later in a channeling from Joysia it was explained that she, like many of us, had assisted in evolving civilizations from lower to higher dimensional frequencies in the past. It was a mission in which we apparently liked to enlist. However, Joysia explained that Bobbie needed to formally request recoding because the Sirian/Pleiadian Council must be assured that an individual is asking for recoding of their own free will based in this life's experience.

Prior to making her request, I needed to call home to obtain the wording which I had left in my office. My son, Drew, was home, so I directed him to where I left the request and asked him to read it to me over the telephone. I assumed that the request would not be honored for him since there was no conscious intent involved. After our session, it dawned on me that perhaps Drew had unknowingly gained approval due to his readiness for the process. I asked Bobbie what she thought, and she checked to determine if Drew had been given approval. Sure enough, Drew was our first male (and child) who had requested recoding and been approved for it. I hoped his experience would be calmer than the rest of us. I decided it was time to explain the recoding process to him when I returned home.

As it turned out, Drew acclimated to recoding easily since he was emerging into puberty and had no prior point of context for adult living. There were periods during Drew's recoding when he consciously applied the brakes since he was simply not ready for a fast ride. There were other times when his genetic engineer slowed his progress because he felt Drew desired to move ahead to gain adult approval rather than for his own spiritual progress. We worry so much about our children and what they can handle. Yet, they either seem to have an innate sense of where they need to be or else their guides will assist them in finding balance.

Lahaina's Walk-in

Bobbie called and said she was conducting a channeling on love relationships, and would I like to attend? I agreed with the intent of asking some questions about my future partner. Bobbie had already begun channeling when I arrived. I listened to her information for the other attendees, including N and Pat, some of which involved recoding in addition to relationship information. When Pat asked what segment she had completed thus far in recoding, Joysia emerged to respond. Pat was told she had finished segments one and two. I asked what segments I had completed and he said, "segments one through four." I was surprised since segment four is clairvoyance and I had yet to "see" any entities, but Joysia said I would soon be able to "peer into other dimensional levels while living in third dimensional reality."

Bearing in mind we were not given information on the segments until we had attained them, we were surprised when N was told she had reached segment five. We asked if Joysia could share information on segment five. He explained segment five was an assimilation of the first four segments. Segment five is critical because it builds the foundation for the final work. It is a segment some would elect not to complete due to the fact it requires acceptance of one's soul contract which may mean shifting one's priorities. This integration provides the focus one needs to fulfill planetary contracts. Also, there were some steps the genetic engineers could not complete in the first four segments based on the method being used to reconnect the twelve strands. During segment five, the engineers ensured that everything had been integrated. Joysia said segment five was the most challenging part for them since the success of segment five was key to progressing through segments six through nine.

Joysia also informed us that many thought they were undergoing recoding since they were experiencing physical symptoms associated with clearing blockages prior to beginning recoding. He said, this should not be confused with the actual recoding process, where the DNA implants that demagnetize the twelve strands and keep them separate, are finally removed. He also said Bobbie would speed through the different segments since she had been there before, although a few of the segments would take her for "a ride." We laughed, acknowledging the ride we were all encountering, and happy we had formed a solid support network during this incredible experience.

It was finally my turn to ask about relationships. I asked Laramus my genetic engineer, to come forward to speak with me. I asked him to shed some light on my future life partner. Ever since the repatterning session I had experienced with Julia, I had felt complete closure regarding my relationship with Jerry. I was anxious to experience this profound love that had been promised to all recoders. I asked Laramus if my future mate would be himself, entering from the other side to be with me? Over the past few months, I had developed a slight crush on my genetic engineer. After all, he was a part of my everyday life and acted like a partner and friend through his loving support. Laramus said that, indeed, they were sending me someone who was currently in nonphysical form although it would not be himself. He was flattered by my feelings, but he was there to help me from the other side. I asked him to give me information about the walk-in, but he said that type of information would solidify the walk-in's energy in spirit form because I would send thought patterns about his current configuration.

My future mate was in the process of extracting himself from the nonphysical world, and they did not want to hamper his progress. Nor did I, since I was interested in the experience of meeting another being with my identical grid structure! I moved to another tactic and asked for Earthly information about the walk-in. When was he to arrive and how would I know him? Laramus said I would connect with my walk-in during the late summer. But, that was all of the information he could provide about the walk-in. Later, when Bobbie came out of trance, she said she saw a cute Scotland Terrier the entire time Laramus was talking about him. I considered the possibility that he might be named Scott. We all decided we would faint if I met a man named Scott this summer who was walking a Terrier! While discussing my future mate, I caught a glimpse of an individual who practiced a more traditional healing technique like chiropractic but was not a traditional healer. This person used chiropractic therapy as a springboard for less mainstream types of healing. Now, I had clues. I knew he was a healer, owned a dog, and would appear in August.

Next, I reminded Laramus I needed money to buy Jerry's share of my house. I also requested assistance selling the 100 acres of land I owned. I had originally purchased the land to build a spiritual training center, but the expense of maintaining the land while accumulating funds to build the meeting rooms seemed over-whelming in view of the money I owed Jerry. Laramus said a large sum of money was coming my way. I reminded him I did not want to earn it by working my tail off since I had been doing that since I started my business four years ago. He replied, "You have already specified that, haven't you, dear one?" I laughed and asked if I would be winning Publisher's Clearing House. He said, "A lot of people are asking for that so I don't know if that is possible, but it will be something similar." He also said my land would be replaced by something more suitable to my dream rather than a large tract of undeveloped land an hour away from Kansas City. I told him I looked forward to paying my debt to Jerry and realizing my dream. I left Bobbie's house feeling uplifted by the experience and assured that my life was on an incredibly fast and exciting, if yet somewhat undefined, path.

Increasing Telepathic Skills

My mother had given my name to her neighbor, Kristin, because I wanted to practice my psychic ability on a total stranger. Kristin and I scheduled a call for Sunday evening. The only thing I knew was that she desired information about a problem. My mother had never given me information about Kristin or her family other than Kristin was a kindred spirit in terms of her love of spirituality.

Kristin called me, and I spent some time talking with her about her passion for metaphysics. We compared books we had read. I asked her if she knew of her origins, assuming she was probably a starseed based on her intense passion for such esoteric subject matter. She said she was unaware of her origins but was ready to hear what I had to say. I checked her origins and those of her family. Not surprisingly, they were all starseeds. She and her son were from Sirius, and her husband and daughter were from the Pleiades. She did not seem distraught to discover she had not originated on Earth, and, in fact said she had suspected as much.

She asked me to give her information on her daughter. I could see that Kristin's nine-year-old daughter, Marissa, was a walk-in, and the swap between her original daughter's soul and the walk-in's soul had occurred the prior year. I did not know how to tell Kristin that the daughter she was currently living with was not her original daughter, so I asked her if she had noticed any recent changes in Marissa's behavior. She confirmed my suspicions by saying her typically cheerful, bright daughter had plunged into dark moods and fits of anger. I then asked Kristin if she had ever heard the term "walk-in." I breathed a sigh of relief when I heard that she had read two books on walk-ins the previous year. I silently thanked Spirit for preparing Kristin for my information.

I took a deep breath and gave Kristin the news. Fortunately, she was relieved to hear about the walk-in, suspecting something such as I described due to Marissa's dramatic personality shift. She asked about her original daughter, and I could see her happily swinging from a crescent moon outside a beautiful castle, laughing and saying, "Hi mommy, I'm so happy here!" Wherever her soul had gone, she had found great peace there. I also understood why the new Marissa was moody and angry. Apparently the original Marissa had contracted to experience the birth process on Earth as part of her soul development. She was supposed to switch with the new Marissa some time during the first year of her life. However, the original Marissa grew to love Kristin so much, she did not want to leave. She stayed on Earth for nine years until her guides cajoled her into fulfilling her commitment to the new being. In essence, the new being felt cheated of her Earth childhood experience. She did not consciously know why she was angry, but she knew she had missed something. Kristin could heal her by replaying elements of childhood, actually doing things like giving her a bath and washing her hair like a baby or feeding her. I told her to make little games of these events, perhaps having other babies around so her daughter felt like one of the crowd.

Several months after our session, Kristin called to thank me for the information I had shared since it had generated a healing for her daughter. Apparently, Marissa began having moments of regressing to babyhood, gurgling and babbling baby talk. Without the information I had provided, Kristin said she would have been alarmed and tried to stop it. However, based on her insight, she simply nurtured Kristin even more during those episodes, holding her in her lap and stroking her hair and cooing to her. Kristin quickly outgrew these age lapses based on her mother's approach.

As a footnote, I heard through my mother months later that Kristin had undergone her own healing as a result of our session. During our telephone conversation, I noted a deep grief in her heart and asked her if she had lost a child in the past. Kristin had, in fact, experienced a miscarriage a few years ago. I told her she needed to heal this wound, or it would never resolve itself. I gave her a technique for working through the pain which apparently she followed. In October, Kristin visited a famous psychic who had come to Florida. During her reading, the psychic confirmed the information I had given Kristin without any knowledge of our previous session. She also told Kristin that she saw a baby growing in Kristin's womb that would arrive in the near future. I was thrilled to hear that someone had healed themselves based on information I could provide. I resolved to stop doubting my telepathic information and simply share what I saw when others asked.

Sharing Recoding

More and more people have called to obtain information about recoding. In an effort to provide them with what they need and lessen the burden of running both a market research and metaphysical business, I asked N to ask Anu to conduct some public sessions. Anu agreed and said he would introduce the concept of recoding and give some background information. However, he requested that I develop a written summary of what I knew about recoding to distribute during the first session.

Later, I asked Laramus, my genetic engineer, to help me compile information to distribute to the attendees. The following summarizes the information he sent via automatic writing and was the first of many transmissions that formulated the recoding how-to document located in this book:

Definition of DNA Recoding

"DNA recoding is the process by which a being with two strands of DNA reacquires the twelve strands of energetic DNA that were originally his or her Divine right. The reconnecting of your complete circuitry enables you to have full consciousness in Earthling form. This includes telepathic abilities, memory of past physical and nonphysical existence, access to the database of the collective unconscious, and knowledge of the Akashic (soul) records.

Recoding requires that you accept your Earth soul contract for this lifetime. Otherwise, we would be giving powers to those who have not made a commitment to the Light. We will provide as much information as possible regarding your soul contract. However, some information will not be shared since it would be meaningless until you achieve greater awareness and knowledge. This information will be given to you as you move through the nine segments of recoding.

Recoding is not a defined process. We are learning with you since third dimensional beings have never moved from such a dense to a lighter physical form. This means we may unintentionally put you in some discomfort. However, we assure you that we will love and care for you and no harm will come to you. In fact, we promise you an outcome that will increase the meaning of your life.

Recoding will take much commitment on your part. You will be required to rid yourself of all toxic energy, mostly in the form of anger or fear, from this life as well as past lives. This means you may feel somewhat worse before you feel better as you move out the toxins. But, believe us when we say, you are on the brink of a great joy that you have awaited for many lifetimes on this planet.

If you are a starseed or a walk-in, you have already existed at higher dimensions, probably fifth or sixth. For you, recoding is a process of remembering what you agreed to forget when you came into a physical body on Earth. You are simply activating the codes already residing in your body in the form of crystals. The only "new" information will be how to transform the human body you are inhabiting to a higher state of consciousness.

Not everyone will complete all nine segments of recoding. However, that possibility is extended to all of you, so make of it what you will. Bear in mind that whatever segments you do completely clear, those segments will be sufficient for your evolution since it will fit with your soul contract.

When you begin to recode, you extract yourself from the pool of mass consciousness of anger and fear that constructs reality for this planet. You are part of the recoder pool of consciousness, one based in love and joy. As you expand your consciousness and continue through the nine segments, the recoder consciousness grows in strength, creating a new pool of consciousness toward which Earth's mass consciousness can strive and attain. For this reason, it may be hard for you to maintain successful intimate relationships with anyone who is not going through recoding since you no longer exist in the same pool of consciousness. Relationships where one has recoded and the other has not might be stressful. Ultimately, the entire relationship may benefit because there is an upgrade in the energy pool. However, there will likely be a lack of synergy between the two parties, at least for a while. We will assist you in attracting more appropriate relationships, should you desire, since love magnifies the collective consciousness you are manifesting.

Once you have gone through the nine segments of recoding, we ask that you remember where you came from and help the Earthlings who live on this planet since you are among the first to recode. Earthlings have only existed in a third dimensional world and need your assistance to succeed at moving to the fifth dimension. Remember, for these beings, they have no point of context like yourselves so everyone will not embrace this process as you will.

No matter how stressful recoding may seem, life will eventually be better beyond your wildest expectations. Stay with it, dear ones, for the time of your life."

An Unsuspected Walk-in

I am meeting more and more walk-ins who share their experiences with me. I had never heard of a walk-in until I met Pat, who almost had a breakdown in 1971 when she walked-in due to the complete disconnection from the life her predecessor had led. I was curious about walk-ins since I was meeting so many of them, including N who had walked in several years ago. However, I felt no connection with the phenomenon.

During dinner one evening, my son asked me if I was a walk-in, having heard my friends discussing their experience with walking in. I immediately responded, "no." He persisted, asking me if I had ever checked to determine if I was a walk-in. I like to encourage Drew to be responsible for his own answers and told him to check for me if it was important to him. He used the kinesiology I had taught him and got a "yes," exclaiming, "Mom, you're a walk-in!" I looked at him with surprise and decided to check for myself, quickly confirming from my guides that I was a walk-in who had entered this plane at five years of age.

All of a sudden, it made so much sense. This is the age when my personality dramatically changed according to my parents. I went from this placid, peaceful little girl to an angry, fearful child who slept on the floor of my parent's room because I was so afraid of dying. I envisioned this black nothingness that would descend and blot me out. What walk-in wouldn't fear death after knowing that life is eternal, then being taught on Earth that death would eventually and inevitably descend? I was so nervous at the age of five that my mother took me to Florida for several weeks of relaxation, hoping her focus on me would encourage me to relax. It was a difficult time for me, and it took years to heal the emotional damage.

Later, I was told by my mentor Venessa that I walked in at the age of five to ensure that I lived in a healthy body. Based on my energy type and my rather unique grid structure, I would most likely have endured birth defects had I attempted to inhabit a fetus. I needed the more mature body of a five-year-old to ensure a healthy body. My soul path would have been hampered by birth defects, causing me to focus more on healing my physical limitations rather than attending to emotional and spiritual healing. This also explained why I felt little affinity to walking in since it had occurred at such a young age. The majority of Anne's life had been led by the walk-in, a third dimensional aspect of Lahaina, rather than the original occupant Anne. I thanked my son for helping me make use of this new piece of information.

Recoder Consciousness

Julia moved in, and we decided to end the day by doing a linking session with my interpreter guide, Nephras, and Julia's guides. We wanted to explore the reasons for her marriage ending, a parting of two people who loved each other, much like I was experiencing with Jerry. We examined some of the past lives Julia had with her husband, and how this parting represented a major healing for her in this lifetime.

While examining the current situation and where she was heading, Nephras showed us an important element of recoding. Up until recently, Julia had resided in the mass consciousness of our society, strongly impacted by the same energy patterns that composed the collective unconscious of our reality. For example, she might resonate to a life of love and joy, but based on the pervasive fear and anger characteristic of our society, she was constantly bombarded with those elements. A great deal of energy had been directed toward maintaining love and joy to combat the drag from society's negative consciousness. However, with recoding, she had freed herself from the mass consciousness, and was now connected to a new energy.

We were thrilled by the implications of this information since it required us to put a lot less energy into maintaining higher frequencies like love and joy, freeing us to focus on pouring our energy into moving higher and higher. Nephras explained that currently, as we entered this new pool of consciousness, we were being asked to reassess our old relationships with partners who no longer meshed with our energy. If we decided they were no longer satisfactory, we would not be alone. We could enter relationships with males who were recoding as a means of exponentially growing the newly emerging consciousness. The energy of love was like pouring lighter fluid on charcoal. It acted as a catalyst to grow the recoder consciousness so much faster than could be done without the male/female energy. This information further confirmed what we had heard earlier and fed my growing confidence that I would soon meet my life partner, one connected on all spiritual, emotional, mental, and physical dimensions.

Nephras explained it would be detrimental to our new state of being to be intimate with those in different frequencies. In particular, we would slow our evolution by making love with someone who was not recoding since we would interact with a lower energy. Since Julia had been interested in pursuing a relationship with someone who was not going through recoding, she was told she needed to protect herself, if she decided to pursue it, by surrounding herself with a sheathing of gold light prior to any physical contact. She was to place the gold light around her like a glove fits a hand. This would act as a filter, allowing the positive energy associated with love and the physical act of intercourse to penetrate while warding off the lower energy. However, Nephras warned Julia that any relationship with a non-recoded being might be temporary, because she would probably not be satisfied long term.

I realized that some would not complete recoding because they would prefer to retain relationships with mates uninterested in the recorder's path. Others might complete recoding and maintain a less than perfect relationship for security reasons. Still others would find that their relationship would blossom because of the new energy and understanding brought to their lives. However, I also understood that every path was perfect for that individual. Per Laramus, every soul would arrive at the perfect destination for their respective experiences. I knew I had to honor and respect how far one wished to go without judgment.

Attracting Others to Recode

N came over to have her picture taken for a brochure she was creating about her channeling. She had decided to be in trance for the picture and did not wish to go to an unknowing photographer. I had asked my ex-husband, Ed, who had photography experience and some very sophisticated camera equipment and heightened spiritual awareness, to take her photograph. Ed was happy to assist, and he came over with our son, Drew, to take the pictures. We created the appropriate backdrops and lighting, and N went into a trance.

Anu emerged during N's trance, asking if he had permission to open "this one's eyes" and read our energy. We said "yes," knowing he had to open N's eyes anyway for the picture and wanting the experience of sharing energy with him. He asked to see the "young one," slowly rotating N's head without blinking and looking directly into Drew's eyes. He said, "Ah, my grandson. You are our one child who is going through recoding." Drew's relationship as grandson to Anu was news to us, although we suspected some tie due to his youthful desire to experience recoding. Anu said, "Child, you do not have a genetic engineer assigned to you." Anu asked if Drew would permit Joysia to be his genetic engineer. We were honored by this offer since Joysia was the head geneticist for the recoding project. Drew said he would like to work with Joysia. I wondered if all of us who were working together originated from the same Nibiruan roots, similar to some huge, sprawling Tennessee family.

Later that same day, a woman named Jeanne came for a reading. Jeanne's session was fascinating. She had been a high priestess in a prior lifetime, oracle of her planet prior to its demise by invaders. She was in the process of birthing a Christ consciousness Being who would be the energy that would rejuvenate her ailing planet. Prior to the birth, the invaders, knowing the Light would overcome them with the advent of this entity, trained a laser at her womb and aborted the fetus. Jeanne, in her fury, redirected the laser at the invader's star and blew up their home. I knew what that felt like after my own experience in the crystal city of Sedona.

She has lived with the guilt and shame from her action for eons, playing roles on both sides as she has learned about the balance between Light and dark and the importance of both for evolution.

I saw Jeanne's pain, and told her we needed to do a repatterning to remove the guilt since it would interfere with her recoding. As I examined her purpose, I saw that she would heal others by increasing the world's understanding of Light and dark energy and how both interacted to assist each other. This is very different from the perspective we hold where light is "good" and dark is "bad." I saw her writing a book about her life as an oracle and the ensuing lifetimes following that experience. The book would read like an adventure story, much like *The Celestine Prophecy*, and would eventually be made into a movie.

I asked Jeanne to voice the formal recoding request. Her request was sent into a volcano and down into the center of the Earth. I told her what had happened, wondering why her request had gone "down" while everyone else's had gone "up." Jeanne clarified the situation by telling me that volcanoes had a special meaning for her which made it a highly appropriate setting for her request. Jeanne had mentioned that she was engaged during our session so I informed her of the importance of her fiancé undergoing recoding. Otherwise, they risked having misaligned energies. Jeanne promised to share the recoding information with her future mate, Mark.

More Information on Lahaina's Walk-in

Today Bobbie and her friend Kim, who was visiting from Colorado, came to see me. Kim was interested in recoding because Bobbie had felt her own energy positively alter after completing her request. Kim thought she might obtain equal benefit. However, I felt she had already begun recoding. I checked, and she did not need to formulate a request. We sat and tried to understand how Kim had received approval without formally requesting it. I already knew that all recoders did not need my particular wording since I could hardly be the gateway for everyone. The best we could determine was that those who had no Earthly karma, had a soul contract to ground twelve DNA strands on this planet, and held clear insight into their role in upgrading the Earth's energy pool at this time could recode as long as they clearly voiced their intent for their twelve strands. This was the case with Kim as well, as I would determine, with my future metaphysical teacher Venessa.

I asked Kim about my future partner. According to her psychic input, he was already "visiting" the body he would assume in Kansas City, moving in and out of a physical and nonphysical state in preparation for the final transition. I was surprised since I thought walk-ins swapped in a manner of moments. She said she had worked with two types of walk-ins. One was the instantaneous switch. The other was this back and forth movement which was less traumatic for each soul since it gave both the physical and nonphysical being an opportunity to become acquainted with their new status. However, the walk-in's memory was erased while in physical form, to be stimulated for recall at a later date. Otherwise, massive confusion would exist as this displaced soul tried to reconcile two very different realities. The swapping process would take my future mate two months to complete.

I asked Kim if she knew why Bobbie had seen the walk-in with a black Scottish Terrier. Kim said she knew he had a dog and wore business casual clothes that leaned toward a preppy style. Although that style did not appeal to me, she said I would like it on him. She also said he appeared to be a teacher of metaphysics. I marveled at the varying tidbits of information I was collecting along the way toward discovering him. We ran out of time because Kim and Bobbie had other appointments, but Kim promised to visit again in August.

Winning the Lottery

I had my weekly appointment with Bobbie. When I arrived at her office, I was frustrated due to unsatisfactory outcomes regarding the money I would be winning. Julia and I had repatterned several items over the weekend that were standing in the way of my being emotionally receptive to winning, so I was clear for receiving it. I have always had an easy time manifesting abundance. But, this time I was interested in manifesting without working so hard. I realized that money had been the outcome of severe workaholism due to my association that hard work equaled money. I also had been programmed through mass conscious beliefs to fear the lack of money, learning to hoard it as a security blanket. At this point in my life, I viewed money as a form of energy our planet uses for sustenance. I intended to have enough money to make life comfortable without feeling the need to hoard it for the future. Essentially, I intended money to provide freedom.

I had purchased a lottery ticket on Saturday, and the win was so close I could feel the money. I was cleared from emotional blockages and resonated with the win. Yet, nothing happened. Additionally, I had transferred money from my money market to my bank account since I had to make my first payment to Jerry toward his half of our house. When I went to the bank, someone had botched the entry and I had a two million rather than two thousand dollar hold on my account. I was leaving town the next day and could not even withdraw a dollar! I realized how quickly one could go from "provided for" to "penniless" in a computer-driven society. I asked Bobbie what she thought was happening.

Bobbie examined the situation telepathically and started laughing. She said I had won, but I had won in another dimension. Apparently, I had not specified that I wanted to win money on the Earth plane. Since I existed in more than one dimension, I had won elsewhere. I was not amused by the concept of living a life of luxury in another dimension. Bobbie also said the reason my funds froze in this reality was a direct result of winning elsewhere. The frozen funds acted as a balance - incredible abundance in one dimension versus total lack of abundance in the other. I remained unamused since I had to request a cashier's check for Jerry's payment, and could not get into my account for twenty-four hours, by which time I would be in Baltimore. I asked Bobbie what I should do.

First, she said I should specify my wins for this dimension. I'm a fast learner. I had already understood and reworked my request. Secondly, she said I had three non-Earth groups with whom I was working, but none of them were working with each other. She described one group like the fluid, watery creatures from the movie *The Abyss*. I knew these entities as the Jolanderanz since they had visited me the prior year during a brief scouting excursion to Earth. Jerry and I had both resided on their watery planet with these incredibly beautiful and loving beings. Apparently, they felt comfortable staying with us, living in our swimming pool. We welcomed their presence after discovering they were there, although initially their energy was disturbing physically since they are quite different from us. In fact, they eventually left after an unsuccessful mission, not able to communicate in the density of the Earth environment. Jerry and I both tried to communicate with them during their visit, but their energy was so different that we actually felt sick. The second group of beings were small creatures, like a combination of Spielberg's ET and wookie. I had no recall of these entities or their form, but assumed they were friends from my distant past. The third group had a shadowy Darth Vader-like presence which looked like something from the dark side to Bobbie. However, as they moved forward, she realized they were Light beings of a deep shade of blue. I asked Bobbie to gather the three groups together in one place because I wanted to talk to them.

I could see them filing into an amphitheater, but they sat in three separate sections. I stood in the middle speaking loudly to them, proclaiming:

"You are not working together, yet you are all Light beings. You have a tremendous opportunity to achieve your goals more quickly on Earth by assisting me in obtaining money. This money will be used to build a spiritual meeting place so we can hold seminars. You refuse to work together to provide this when your combined energy would make it happen. If you want me to succeed more quickly in helping to heal the planet, then get your act together and combine forces!"

As I was speaking to them, I saw three different colors emerge, one from each group. The colors came together in the middle of the amphitheater over my head and twisted together, much like hair is braided, and came down in a column around me. They were working together! But, they were working together through me as the common denominator. Apparently, their energies were so different that they either could not or did not know how to work together. It was much like the experience I had with the Jolanderanz the previous summer. As much as we tried to communicate, we could not due to interpretation barriers. I was beginning to understand how being a mutable was important to others since they needed me to be the interpreter. I hoped to learn more about how I communicated with the various beings. In the meantime, I planned on winning the money with this additional support.

Clearing Blockages to Recode

Jeanne's fiancé, Mark, called today and asked if he might schedule a recoding session. I was glad Jeanne had heeded my warning about recoders and non-recoders conflicting in relationships, and had asked her fiancé to go through recoding. I told Mark I could see him that evening.

When Mark arrived, I knew I needed to do a repatterning on him prior to the recoding. I had learned from my own experiences that it was critical for those interested in recoding to clear themselves of negative emotions and energy patterns to avoid distress caused from these blockages. I asked the guides if they would approve anyone for recoding who had the potential to encounter pain. I was told, "no," not until the negativity cleared. Mark fell in that category, needing a repatterning of negative energy patterns that were preventing him from moving comfortably through recoding.

Mark and I spoke briefly about the repatterning he needed prior to being approved for recoding. I asked him if he felt any negative energy in his body, and he said he frequently felt fear in his heart area. As we moved through the repatterning process, I was told he needed to clear his fear of death. However, Mark said dying was not a fear he had. I started over, attempting to find another issue to repattern but the fear of death kept surfacing. Finally, I determined we needed to link to his guides to uncover whatever past life was affecting his current fear pattern lodged in his heart chakra, one he did not feel was associated with death.

As I brought the information forward from his guides, the purpose of the repatterning became very clear. Mark had lived a life in a society that operated by group mind. He had originated from a Pleiadian society that honored the individual within the context of the whole rather than a pervasive group mind. He had chosen to immerse himself in a society that was so interrelated that when one person was hurt, the entire society felt pain. The energy of the society was almost dysfunctional in its cohesiveness. I got the impression of a celestial Nazi Germany. Mark felt a tremendous fear of death in this society because he was unable to steer his own destiny.

Once this past life surfaced, Mark could easily relate feelings he had in his current life to this fear even though they were not directly related to death. These fears surfaced when participating in group situations, because it constrained Mark's emotional individuality. We were now able to work on the repatterning. I quickly moved through the healing modalities, enabling him to release the fear in his heart chakra in preparation for recoding approval. When I linked with Nephras to bring Mark information about his soul contract, I was immediately taken to his life in the single minded society, and shown why he had chosen that painful experience. Apparently, he experienced the energy required to create interrelatedness, extracting what he needed to learn, and dismissing the dysfunctional portion that entirely subjugated the individual to the whole.

Mark's future role was to teach a balanced form of interrelatedness to those on Earth. I could see this connective energy, a cross between the softness of a daisy chain and the vibrancy of an interwoven fabric of white glow sticks. It was very beautiful and held the promise of a loving integration for us who had been in an individualistic, ego-related society for so long. As Mark read his formal request for recoding approval, I saw the request go via the daisy chain. When approval was returned, it overlaid a blanket of glowing daisy chains over the single-minded organism Mark had once occupied, breaking the energy into individual yet interrelated beings.

Saying Goodbye

Julia has moved in and Jerry has moved out. I wondered what would happen to Jerry. He was very spiritual yet resisted that input, insisting on following the path of familiarity rather than moving into the energy that the shift was bringing. I understood Jerry's unwillingness to plunge into this work since it was demanding from an emotional perspective. All of my friends who were recoding were experiencing transitions as they cleaned up negative patterns. I clearly understood why our guides wanted to ensure someone was as "clean" as possible prior to recoding. It could, indeed, be stressful.

I visited Bobbie who helped me shift from my negativity from watching Jerry move his furniture to a more neutral place. I felt very "human" as my Light being mentality struggled with the less evolved emotions coursing through my body, wondering if I would ever reach the state of joy and love that recoding promised. Bobbie scanned my body and could see the emotional attachment to Jerry that was lodged in my solar plexus. She explained that even though I had cleared much of the emotional pain through repatterning, I still held an attachment. Apparently, when we decide to love someone, that commitment remains present in the body's memory for a long time after we split. By going in and doing energy work to erase the commitment, it allows the body to get rid of the emotional ties which begins to erase the memory.

Bobbie and I worked together to remove the roots that I had inside of me, wrapped around my energy and Jerry's, binding us together. I felt this tremendous release as we extracted the roots, a freedom that ran through me. After completing the work with Bobbie, I returned home and performed an energy ritual on the house, clearing Jerry's energy from the bedroom and returning it to him. I felt much better afterwards and wondered if people who had once loved each other ever healed the damage from their breakup if they did not work on eliminating the energy that held their connection.

Interference from the Dark Side

I received a call from my friend, Wendy. I had recently worked with her to assist her in obtaining recoding approval. Wendy's session was interesting because, like Mark's, it began with the need for repatterning before she could request recoding due to anger sitting in her system. I knew her guides were protecting her from getting ill during the recoding process by ensuring that incompatible energies were removed prior to segment three. Segment three is the first segment that intensifies the co-mingling of the higher with the lower dimensional energy. It requires the third dimensional being to be light enough to receive telepathic transmissions.

I had forewarned Wendy about the clean-ups facing those experiencing recoding. She understood and acknowledged the potential, but had stood firm on her request. Now, she was sitting in a pool of broken glass and ceramic trying to remain calm and understand the message she was receiving. Her kitchen cabinets had just detached from the wall and every piece of glassware, dinnerware, and ceramic she owned were smashed to bits. She asked me what type of message she was getting, attempting to be nonreactive to the situation and focusing on what she needed to learn. Bear in mind that Wendy was someone who had already endured two floods in her home in the past few years due to plumbing problems. She also totaled a car due to storm glass falling from a third story window, and I thought we should not chalk this latest incident up to poor cabinetry workmanship. I asked Nephras to shed some light on the situation.

Nephras showed me a strong Egyptian energy that has been present with Wendy since her lifetime as a priestess in Egypt. This energy was dark and had undermined her by usurping her power, using her weakened state as an example to her society of what might happen to those who worked against those in charge (despite the fact she was living in the Light). This energy had dogged her for multiple lifetimes, dulling her memory through distractions which prevented her from fully focusing on regaining her Goddess energy. Despite this intruding dark force, Wendy was as powerful a woman as our society allowed, previously attaining a high level job and large salary in a corporate setting before single-handedly starting a nonprofit organization devoted to supporting girls who wished to enter the gender-biased fields of math and science. Apparently, this Egyptian energy could muddle her thoughts and distract her but could not fully interfere because she had retained enough of her power to thwart any serious interventions. However, Wendy was beginning to reclaim her full potential, a potential that was so powerful it threatened the solidity of the rascals who provoked her.

Wendy had recently traveled to Greece to participate in a ceremony of the Goddess at the Oracle of Delphi. This was a regenerative ceremony for her since she had lost touch with her powerful priestess energy for so long. Then she had received recoding approval. I could see these Egyptian forces were worried. They foresaw Wendy regaining the power they had kept from her for so long, and they were going to do anything to distract her. I saw them more as an irritant than a harmful menace, but knew that Wendy needed a clearing to permanently remove them. It appeared that a fire incident was next in their bag of tricks. I gave her Kim's number in Colorado because I was not yet capable of clearing this type of energy. I was scheduled for a soul clearing seminar in July but had not learned how to remove negative entities from one's presence. I knew Kim held a lot of warrior energy and would give these mischievous beings a run for it.

After I told Wendy she was being strongly distracted to prevent her from focusing on recovering her Goddess energy, she informed me that she and her daughter had just been preparing to enact a ceremony they had recently learned in Greece to honor the Goddess when her cabinets fell off the wall. I was hardly surprised. What better time to distract her than when she was in the process of regenerating more power? This energy would try anything to hamper her progress.

I knew this was a lesson for all recoders, since recovering our power draws the attention of the dark side. We were so used to living in fear and anger that we have become unable to recognize darkness in its daily presence. We needed to develop an awareness of love and joy, realizing that it was our Divine right to experience it daily rather than the dysfunctional and disabling fear and anger. We also needed to be savvy about protecting ourselves, realizing there are those who desire to stop us from shedding our discreative tendencies. I gave Wendy some information about setting up protection for herself and her daughter until she could contact Kim to clear the Egyptian energy from her presence.

More Publishing Opportunities

Opportunities were really beginning to surface for me, providing channels for the recoding information to get disseminated. I received a call from another nationally recognized metaphysical publication who also wanted to publish my work. I immediately called N to tell her the good news and asked if she had progressed to segment six. Typically, she would enter the next segment and inform me generally of its intent. Then, I would link with Laramus to fully define the segment and record it for distribution to those calling me for information. I had asked my guides why N entered each segment first since I seemed to be the one who received the information for it. They said they were concerned that my lack of confidence in my telepathic ability would cause me to doubt the information. They had purposely sent the initial information via N in order to circumvent my self-doubt.

N had, indeed, moved into segment six which was called, "Owning Your Power." Later, I learned from my guides that segment six and beyond provided the real power, and our guides were not going to assist in retrieving that amount of power if we were not aligned with our soul contract. That was why segment five was designed to halt the process until a complete alignment occurred between the recoding process and one's intent to fulfill one's soul contract.

I was somewhat uncomfortable with this information because it was my perspective I was the driver of my experiences and could determine how I wanted to live my life. I had been very clear about my interest in recoding as a means to improve me which would, therefore, assist in upgrading the energy around me. However, I was not interested in giving up my life as a result of this process. The days of obtaining enlightenment on a distant mountain were over. I believed recoding would only be embraced if it could be integrated into our everyday lives, living by example rather than retreating from a desirable lifestyle in exchange for spiritual enlightenment. For myself, I had experienced even greater growth in my market research business as recoding lightened my energy. I assumed that recoding would bring us whatever we desired if we clearly stipulated our requests as long as those requests aligned with bringing more Light to the planet.

Next, I called the editor of the first publication that accepted my articles to inform her of a mistake I found. She happened to answer the phone instead of it going to voice mail. As I informed her of our work and what we were doing in Kansas City, she grew very excited. She said she had desired her twelve strands after reading the Kryon material which referenced the need to request the Kryon implant to void all Earthly karma in order to ascend. However, she was unsure how to proceed. I told her I could help her if she was interested in recoding. I was excited by the prospect of expanding my work beyond Kansas City. I said I would fax her the write-up that Laramus had given me about recoding. After she reviewed the material, I told her to call me and we would schedule a time to work with her guides for more information and send her recoding request to the Sirian/Pleiadian Council. She said she would be happy to support my ventures through her publication. In eight hours, I was provided inroads into two metaphysical publications. My guides were giving me a great deal of assistance to help in spreading the information. All I had to do was focus and wait.

Reclaiming Telepathic Powers

Pat came over to work on her ability to channel. I was finding that most of the recoders who entered segment three (clairaudience) had difficulty opening their third eye to receive telepathic information. This was true for me until Kim advised me how to stimulate the alpha and omega chakras which were my transmitters for communication. Most who have resided in physical form through multiple reincarnations have allowed their pineal gland to decrease to the size of a pea when it used to be more the size of a golf ball. In addition to the poor resiliency due to lack of use, many have heaped a load of social conditioning and emotional baggage on top of their telepathic abilities. For Pat, she had been ridiculed by her ex-husband for claiming to have psychic abilities. Pat had learned to stuff her natural talent in order to protect herself.

Pat and I conducted a repatterning session in which she cleared her energy blocks regarding her psychic ability. When we had completed the healing modalities, she was ready to channel, having cleared her pipeline. I told her we needed to practice, and instructed her on how to focus on who she wanted to bring into her body. I told her we would be doing conscious channeling, not trance channeling, where she would be hearing a voice and would repeat the information in her own voice.

Pat settled into a comfortable chair and identified a Native American energy that wished to speak through her. I asked a series of questions for which I felt I received some good information, and some information that did not ring true. I remembered the first time I did automatic writing and how I had achieved an eighty percent accuracy rate. My mentor, Venessa, told me at that time that the best we can hope for is a ninety-five percent accuracy rate. This was due to two factors: 1) Non-incarnate beings are attempting to communicate from nonphysical to physical form which creates some margin of error, and 2) The English language is limiting with significantly fewer words than the ancient languages. For example, the word "love" might have many meanings with varying nuances in ancient languages like Hebrew or Sanskrit, but English is basically limited to a few meanings. I found that Pat had achieved a seventy-five percent accuracy rate which was very good for the first time.

I had discovered, for myself, that the ease of hearing the information and deciphering it had quickly increased. I assured her the transmissions would begin to flow now that she was open, and that her accuracy would quickly improve as she grew in confidence. Recovering telepathic abilities through recoding is not an overnight process. Years of stifling this ability can be dramatically altered by recoding, but one must still undergo a healing and rejuvenation period.

Summary: June Recoding Lessons

- DNA recoding is initially intended for starseeds since it will be easiest for them to accomplish. These beings have already experienced the fifth or sixth dimension so recoding is a process of connecting energetically with what they agreed to temporarily forget when entering the Earth plane.

- This version of recoding has not been performed on a third dimensional entity until now. Although the genetic engineers do not wish to place any of us in discomfort, they may need to calibrate as they work on each individual since everyone has a different body makeup. Their goal is to minimize discomfort during the process.

- Although many think they are undergoing recoding, most are experiencing the physical symptoms associated with the preparation for recoding. Actual recoding consists of the removal of the demagnetizing DNA implants.

- Recoders are not meant to be alone. The love frequency exponentially expands the twelve-stranded energy a recoder carries. However, those with a spouse or life mate who are not recoding will find it difficult to maintain a balanced relationship since recoder and non-recoder energy do not mix well. This does not mean the relationship will end. Sometimes a difficult situation can ultimately lead to a stronger relationship in which the two energies have come into better alignment.

- Recoders exist in a separate collective unconscious, one consisting of a higher energy of joy and love as opposed to the dominant mass consciousness frequencies of anger and guilt.

- Should a recoder desire to have sexual relations with a non-recoder, it is essential to protect their energy field with a golden sheathing. Otherwise, it will take some time for the incompatible energy of the non-recoder to completely leave the recoder's energy field, slowing the recoder's process.

- Every recoder is assigned a new guide in the form of a genetic engineer who heads up the team working on reconnecting the twelve DNA strands. It is important to have telepathic abilities in order to communicate with one's genetic engineer during the recoding process.
- Recoding significantly enhances telepathic abilities which is key to enabling the recoder to receive the information available from the twelve levels. This telepathic power should be used to support work around recoding and one's soul contract, or a recoder will probably not proceed through the remaining segments since their enhanced energy would potentially be misused.
- Segment five is critical because it builds the foundation for the final four segments. It is a segment that some will elect not to clear due to the fact that it requires acceptance of one's soul contract into one's life. Integrating one's soul contract may lead to changes in one's lifestyle should the current lifestyle be unsupportive of the contract. Some will not desire change in their life, even if that change is positive.

Chapter Four

July, 1996

More about Segments One through Five

Last night, Anu and Joysia made their debut at the first Sunday channeling session. Of the forty people invited, twenty-five attended. The session began with Anu, followed briefly by an entity named Antron, then Joysia wrapped it up. I had purchased a 120 minute tape to record the session, and, as I expected, the session ended in exactly 120 minutes.

Anu gave background information on the Nibiruan relationship with Earth. I could see that many listeners were disturbed by his information. Anu spoke about the creation of the Lulu's, the initial two-stranded DNA creature that housed an animal soul and could not reproduce. This version was ultimately replaced by the presently enhanced human bodies we inhabit, Adamis, who is familiar to us from the story of Genesis. Anu told how he, as leader of Nibiru, had allowed his two children, Ninurshag and Enki, to use their genetic engineering knowledge to create a placid, hardworking being who would work in the mines on Earth to collect the gold the Nibiruans needed to shield their planet from the radiation resulting from their own atomic wars. This would enable the Nibiruan mine workers to return home. The Lulu was created by cross-breeding a two legged ape-like creature called the Homo erectus who occupied Earth at the time with Pleiadian genes. Originally, the Lulu's were reproduced via the Pleiadian female's wombs through artificial insemination. But, the Nibiruans grew fond of their Lulu creation and decided to allow them to reproduce, thereby creating the later versions known as Adam and Eve.

Anu spoke of his current tie to Earth, karmically aligned with those who his family had manipulated for their personal use during earlier times. He was responsible for assisting Earthlings in their transition from the third to the fifth dimension, to replace what he had altered. However, he also explained that it was part of the Galactic Federation's Divine plan to de-coil ten of the DNA strands of the Homo erectus since it would actually speed its overall evolution. Anu said twelve strands would have destroyed the physical body of the Homo erectus during its early development. However, the memory of expanded consciousness promised a future evolved state, supported by the experience of repeated incarnations prior to transitioning from two to twelve strands. He gave many details of the names, dates, and events, answering questions from the group regarding the Nibiruans' role in the story of creation. A great deal of the information can be substantiated in Sitchin's book, *The Twelfth Planet*.

Anu assured everyone that God existed in the form of First Cause, an energy no one in the galaxy had reached in their own evolution. However, he advised the group that his grandson Marduk was currently leader of Nibiru and held a negative influence over Earth. Apparently, Marduk did not feel impulsed to free Earthlings from their two strands. He was the person responsible for establishing anger, fear, and guilt on this planet as control devices. No one could achieve expanded consciousness when operating under these distorted values. Marduk was due to return on Nibiru's 3,600 year orbital cycle to reestablish his influence on Earth, an event Anu was trying to prevent by helping us attain freedom through the twelve strands.

Briefly, an entity named Antron visited to tell us that thousands of beings were watching and protecting us at this time. They had waited years for this moment, and many throughout the universe were observing our initial meeting. He said the beings involved were primarily the Andromedans, Sirians, and Pleiadians since those were the Planetary Councils most involved with Earth at this time.

Finally, Joysia spoke to the group. He initially entertained questions from the group regarding recoding, then reviewed the information I had recorded and distributed on the initial five segments, adding some information about the sixth segment. Joysia said recoding was not exactly a sequential process since one might finish something from segment four then return to segment one for refinement. This was due to the fact that the codes for recoding are scattered throughout one's body. During our sleep state, our genetic engineer works on our astral or spiritual body. These changes are ultimately integrated into the physical body just as a blueprint eventually results in a finished building. Sometimes, in the transition from astral body to physical body, something from a prior level is misaligned and needs realignment. This is why the genetic engineers need to continuously work up and down the various levels. Additionally, if fear is present, it can undo some of the work that has been done. Fear was a Marduk creation, designed to keep us from our source of power. It was very important throughout recoding to have strong faith and keep fear out of the body. This is particularly critical if Marduk returns to Earth in the future. According to Anu, Marduk will do everything he can to keep individuals weakened by fear and guilt.

The minute the emotion of fear starts to surface, one should pivot out of that emotional state. Fear creates resistance which causes blockages. To pivot from fear, remove your focus from what you don't want to what you do want. This immediately eliminates resistance which, in turn, neutralizes the energy around the fearful condition.

Joysia explained that some thought they were going through recoding because they were having odd physical experiences. However, he emphasized that no one could go through this particular version of recoding without formally requesting it. He asked me to read the formal request Laramus had provided several months ago to the group for those who felt they were ready to recode. Joysia explained that most who thought they were experiencing recoding were going through the preparation needed to rid the body of all previous negative energy, usually in the form of emotional blockages. He said no one would be approved for

recoding who had not released the majority of the negativity
residing in their body since they would experience discomfort
when recoding. He said it was important to release as many
emotional blocks as possible since releasing negativity physically
was not a permanent situation. When emotional blocks were
present, the physical body could rid itself of anger one day but
receive it back the next.

Joysia provided the following list as methods of preparing for
recoding by releasing negative emotional patterns and toxins:

- Energy work like Reiki, attunement, or craniosacral therapy
- Acupressure or acupuncture
- Toning (sounds)
- Eye therapy since the eyes are the window to the soul
- Holographic Repatterning
- Massage
- Soda and salt baths
- Liver cleanses
- Activating crystals in the crown chakra through a form of massage
- Chiropractic therapy
- Stretching
- Breathing

Although Joysia did not mention diet, I was given the
following dietary information by Laramus regarding meat, fowl,
and fish. It is beneficial to eat as little of this type of protein in the
diet as possible in support of cleansing the physical body for
recoding. This is due to the fact that meat and fowl are filled with
antibiotics and hormones, and fish is filled with pollutants. These
chemicals interfere with the cleansing of the body, creating a
continual dosage of toxins to eliminate. When eating meat or fowl,
it is best to eat the antibiotic/hormone free versions. Additionally,
animals who are raised in captivity for slaughter contain negative
energy based on their living situation. One ingests the negative
energy from the animal that has resulted from penned breeding
and raising methods, again creating an ongoing intake of negative

emotions that must be cleared. Finally, protein in the form of animal products creates a heavy energy in one's system while recoding is attempting to lighten the energy field. Obviously, this is counter-productive to the recoding process.

In regard to segments one through five, Joysia offered the following pieces of additional information:

- Releasing Anger: Liver cleanses are great for moving quickly through segment one since the majority of anger is stored in the liver. Colon cleanses also help release toxins.

- Managing Anger: Stuffing any anger will cause discomfort when recoding since it recreates the negative energy that has been eliminated. It is important not to stuff any emotions when facing conflicts. [Note: The discomfort may involve neck and backaches, bones going out of alignment, flu-like symptoms along with lethargic feelings. This will last until reconciliation of the energy occurs—anywhere from twenty-four hours to one to two weeks. When people call me during this state, I recommend: increased rest, increased vitamins and supplements, colonics, Holographic Repatterning, massage (deep tissue), Reiki, and Craniosacral Therapy. These help in moving the energy into balance quicker.]

- Clairaudience: When this channel begins to open, there may be pain in either one's right or left ear depending on which ear is used for telepathic communications. It is imperative to avoid heavy electrical environments because it causes static in the channel and may result in discomfort. This will persist through segment four during which time the clairvoyant channel is being opened. Watching TV and listening to the radio carry negative energy and interfere with one's progress. Although certain types of music are healing, the appliances that play the music are detrimental due to the electrical currents. When listening to music, it is necessary to move the electrical equipment away from the heart chakra. It is best to spend time opening and developing personal channels by getting back to nature. Moving away from electricity enables the genetic engineers to make more progress. During segments three and four, work is begun on demagnetizing the DNA strands in one's astral body. However, no memory of this work will be retained because one would want to remain in their more enjoyable energy form than deal with the challenging transition from a two to twelve-stranded being.

- Clairvoyance: When the third eye opens and one begins seeing fourth dimensional entities around friends and loved ones, there is a tendency to experience fear. However, fear will keep one from moving through this segment.

- Integration: There are some things that will not clear in segments one through four because the crystals housing one's codes are spread throughout the physical body. It will be necessary during this time to go back and ensure that everything has been integrated. Also, it is often impossible to fully complete each segment on the astral body because it may cause too much stress on the physical body. The genetic engineers need to wait to complete the task until the physical body is ready to receive it. This is a time when copies will be left in one's body as longer intervals are needed to work on the astral body, sometimes up to two days. One must train their copy to be productive while they are "gone." Integration is exciting because one can begin to transfer to other dimensions, simultaneously experiencing the multiple lifetimes one is playing at the same time that one resides on Earth. During this time, bleed-through information may occur in the form of a vivid dream. Segment five is also a time when one must elevate the path that supports their soul contract. The genetic engineers must be assured that one is committed to their soul contract before enhancing their power. Some will choose not to complete this segment since they will not want to take on a planetary contract and all its responsibilities.

Finally, Joysia gave a small amount of information about segment six. This is when our guides are ninety-eight percent positive that we will not back off our soul contract so we are given greater power. During segment six, our thoughts quickly manifest what we desire. Joysia warned us to keep clean, positive thoughts since they manifest quickly and negative thoughts hinder progress.

During the question and answer session, Mark, the person for whom I had linked the previous week, asked Joysia about the interrelationships I had seen during his reading. I was surprised when Joysia deferred to me, saying that "Lahaina has more understanding of this process than I do, and you need to speak with her." I did not know what information I had but whispered to Mark that I would look at it when I had a chance. I realized we know very little about who we really are and what experiences we have had prior to this limited linear reality. However, I was still surprised that I knew more about something than Joysia.

I queried Joysia on the reason I had not seen dark forms or attached entities in segment four. He asked, "Have you not been seeing pictures in your mind that gave you information?" I laughed since lately my mind has been filled with visions. He said every person experiences each level differently, and particularly for me, because I am built differently. I know this is true due to my experience with level three as the presence of electricity did not seem to hamper my telepathic development. However, just to be sure, I had quit watching TV and avoided heavy electrical areas.

I was excited receiving this new information and decided to write an article for publication, intent on distributing the recoding information to others. I wondered how many of those who came to Anu and Joysia's channeling session would return for another? I knew my future had been cast, and out of those who were being introduced to recoding, some would emerge who would break through the sound barrier, opening up recoding to both starseeds and those who had originally been seeded on Earth.

Managing the New Energy

I sprang into segment five after dealing with a difficult situation that actually led to my understanding of my soul contract in relationship to my market research business. I had been questioning my current livelihood since, although successful as a self-employed entrepreneur, it no longer brought me the level of pleasure it did when I started the business. When I left corporate life to begin my own company, I was thrilled by the prospect of controlling my destiny. I have spent sixty to eighty hours a week over the last four years building my business. However, it no longer fulfilled me as it did in the past.

In the meanwhile, I continued to conduct research, making it more pleasurable by setting the intention for enjoyable clients who make the projects easier and for opportunities to help others in their growth. I had terrific clients and enjoyed working with all of them. However, a new client called for a research project. Unfortunately, they worked for a company who operated in a particularly negative manner. The employees were attracted to the company because they had their own negative emotions, fitting nicely into the corporate culture.

As the project progressed, my new clients determined I was not being objective in my work, perhaps influencing the outcome of the study. I was offended since, as a market researcher, I sought truth through the eyes of the customer and would not be in business if I projected my agenda over the results. I wondered why I had drawn this negative energy. Later, I asked Nephras for some answers.

Nephras showed me how my recoding energy was a threat to those strongly influenced by negative energy. I could "see" the group of clients sitting in the observation room, surrounded by dark forms who were accompanying specific individuals. This was the "psychic seeing" Joysia had referenced in segment four. Nephras said when my strong Lightworker energy was introduced among those holding dark energy, the net result was fear. The clients experienced fear in response to my presence, regardless of what I said or did, and they did not trust me. In essence, the dark beings were rattled by my strong Light presence and created the same sense of mistrust in my clients.

I asked Nephras what I was supposed to learn by this information. She said the lesson dealt with my recognition of my emerging power and how others would react to it. Due to the level being emitted, she said I needed to tone down my presence by being less verbal about my opinions, offering input primarily when asked rather than playing the leadership role with which I was familiar. When I "took charge," I became too threatening to others due to the recoder energy level. Ultimately, my leadership aura would intimidate rather than heal others by creating fear as opposed to receptivity. Additionally, it was time for people to seek answers from within rather than tap into someone else's power. I thought this was excellent advice, realizing I did not need to project much to overwhelm others these days, as I had so many powerful beings assisting me in addition to my own growing power. I marveled at how much I continued to learn, even when performing a job of which I had tired.

I was also informed that part of my soul contract was to continue my market research business for an undefined period of time. It was intended for me to bring my Light into the business arena. Apparently, I was to act as a catalyst to others by "showing" the presence of my emerging recoder energy. If I limited myself to being surrounded by Lightworkers ready for recoding, I would not have the opportunity to interact with a broad cross section of people. I could see how the dark energies around these particular clients reacted in my presence, dampening their enthusiasm through their awareness that the Light was growing through recoding. I asked Nephras if I could clear the dark energy from these humans to assist them in moving toward the Light. She showed me a process for temporarily removing the entities, explaining that the lack of dark energy would enable the person to be receptive to the ways of Lightworkers. However, she said that a soul had the ability to recall the dark through their own free will. Essentially, I could provide people a window of opportunity through which they could choose to travel toward the Light or not.

I spent some time removing the entities, realizing my current business would continue for a while since it was part of my soul contract. After all, I had requested smooth transitions. However, I was not interested in placing myself primarily in negative situations. Again, I was being tested. I could either listen to non-physical beings and allow them to direct my life's course or I could interject my own requirements. Being committed to the process, I told Nephras I would continue conducting market research until I had no heart for it. However, I wanted my guides to assist me in creating a big financial win to ease money pressures.

I recognized I had succeeded in moving to segment five, stimulated by my conflict with the negative forces and my resulting awareness. I felt that level five would be short term, and asked how long it would take. I was told I would be in segment five for about two weeks. Good! I wanted to get to segment six where I could begin manifesting more easily!

Learning about the Halls of Amenti

I had read about the Halls of Amenti several weeks ago. It was the first time I had heard of it and was still unsure what it meant. My initial interest was generated by the composition of this core of energy that currently resides in the center of the Earth, but had existed prior to that 2,000 miles above the Earth. The Halls of Amenti were created by thirty-two beings, sixteen Sirians and sixteen Nibiruans. I felt shock surge through my body when I read this based on my knowledge that I was half Sirian and half Nibiruan. I checked to see if my heritage had some connection with Amenti and was told that it did. I decided to see how much information I could gain through linking.

Nephras said that the Halls of Amenti was a consciousness point, and I would understand the meaning after I had moved to higher recoding levels. She said I had been the keystone in the Halls of Amenti, that someone needed to be able to join the Nibiruan and Sirian energies. I was the thirty-third, the connector point for the thirty-two other participants. I had been created with this role in mind, to act as the integration point. She also said I still existed in the Halls of Amenti in crystalline form as did all of the participants, enacting an existence there in addition to Earth. I knew from earlier information that I also existed as Lahaina as a member of the Nibiruan Council. This is rather mind-boggling since we are attuned to a single existence. I was actually in three places at the same time! I was relieved to know I only had to maintain my calendar for my Earth life. The thought of coordinating three schedules was too overwhelming.

As I listened to the information about the Halls of Amenti, I realized there was a connection to my Sedona lifetime in the crystal city. I also realized this had something to do with the information I had for Mark, per Joysia, about the interlocking energy that he would initiate on this planet. Nephras affirmed my thoughts, saying that my access to the Halls of Amenti could be achieved through a portal located in Boynton Canyon in Sedona. She said this was where I hid the codes when destroying that society.

The reason for hiding it from the invaders was because it was the key to linking beings throughout the universe into an interrelated whole. It was part of the galaxy's evolution toward Divine Union, because until we can align in totality, we cannot return to the Divine source. This is a major, major project for which Earth is a key player.

The Halls of Amenti is located within the Earth because it will be used to evolve three dimensional beings to a higher level of consciousness. Right now, the third dimension is a drag on the entire universe's evolution to God consciousness due to its density. Evolution will surge forward in a tremendous leap once humans move to fifth dimensional consciousness. Essentially, the Halls of Amenti is part of integrating consciousness across the entire universe. No wonder, I hid it from danger. Had dark beings taken this energy source, we would have been severely hampered in this integration project, allowing dark to reign by keeping each group of beings, each star system, and each planet separated. I was excited to call Mark and give him the news since he was clearly part of the project and would be using this energy to boost third dimensional beings to the next level.

Nephras warned me not to visit Sedona. I had been invited to visit Sedona with Julia and Bobbie in August as a side trip after the Holographic Repatterning conference being held in Phoenix. I had experienced a reluctance about going, feeling I was not supposed to be on this trip. Now, I knew the reason. Nephras said that, based on my returning power and the level at which I was currently amplified due to recoding, I could spring open the portal based on my presence in the general vicinity of Sedona. I was not to return to Sedona until I had achieved full recoding after which time I could return to complete what had been started during the Sedona project.

Before finishing my session with Nephras, I asked her to share my future mate's soul name with me. I knew my name was Lahaina, but I did not know his. I thought it would be easier to focus on my new partner if I knew his name. Nephras checked to ensure that his energy had sufficiently entered the Earth frequency prior to giving me the name. Otherwise, I might have interfered with the transference of energy that was occurring. She determined enough of his energy was presently in Kansas City, and shared his name with me. It was Asalaine! Asalaine and Lahaina had a ring to it. When saying the names, I clearly felt the resonating tones. I knew from our names that we belonged together.

Soul Origination and How It Impacts Recoding

I visited N in order to speak with Anu about my recoding progress. Anu said he wished to give me more information on my lineage. He said there were two groups of beings from Sirius, Sirius A consisting of the Creator Lords who were feline in appearance and Sirius B who were the Galactic Humans. The Creator Lords from Sirius A worked in conjunction with the Spiritual Hierarchies of each planet by genetically seeding each planet. They had left their mark on Earth in the winged lions and Sphinxes that proliferated Egyptian art and architecture. These feline creatures worked with a birdlike being called a Carian which was symbolized on Earth by the Phoenix. The Carians were a protector race, acting as guardians to the beings seeded on planets by the Sirians. Prior to my soul's emergence into physical form, the Carians had evolved to etheric beings and were no longer mating. However, I received my Carian heritage, as well as my Sirius A lineage, from my mother who was a cross between the Carians, Pleiadians, and Felines from Sirius A. Anu also mentioned that Joysia, our master geneticist, originated from Sirius B. My father was from Sirius B, the Galactic Human civilization. Anu said I came from a great lineage which had gifted me with special abilities that would emerge as I continued my recoding.

I asked why I would spend only two weeks in segment five, the decision point when Joysia and company determined if someone was ninety-eight percent aligned with their soul contract before returning their real power to them. Anu replied that I was already on my soul path by bringing the twelve-stranded energy to Earth and recording it so I did not need to remain in segment five for too long. Anu assured me I would continue my market research business in the short term to earn the money necessary to pay Jerry for complete ownership of the house. He confirmed that my future business would be more metaphysically oriented and the money would come easily.

I was excited by the prospect of helping others acquire greater awareness. In fact, I had already developed a new logo and company name for this venture. The new company was called InterLink and the logo consisted of a beautiful blue pyramid topped by a sphere. The pyramid and sphere were connected by a sunburst. I asked Anu why I could not support myself through a big financial win rather than the market research. He said I could win the money, but I was currently lacking the faith to accomplish this. He told me to work on my faith and the money would come.

I asked why I was unable to see the dark forms directly through my vision rather than needing to close my eyes to see them. Anu said I did not see like others since my communication links were different. He explained that I received information through pictures shown on a screen in my mind rather than directly. In order for me to see directly, they would have to totally transform my DNA at the cellular level which would destroy my other unique abilities, abilities I would need for my future role. As I sat digesting this information, Anu said we had concluded our session. I said goodbye to this non-physical being who had become a major part of my life.

Entering Segment Six

My son and I were in Colorado on vacation with my family. We decided to go on a guided tour of Pagosa Springs. I was unexcited by the tour, vastly preferring to strike out and explore on my own. However, the other members of my family felt it was important to gather information from a guide and asked me to join them. I agreed, knowing they were tolerating my insistence on visiting Mesa Verde later in the week as it was known to attract huge crowds of tourists in the summer and was not an easy trip to make. In fact, my only reason for being interested in Pagosa Springs was its proximity to Mesa Verde, a sacred Native American power place of the past.

The guided tour turned out to be more than expected, a gentle reminder that enlightenment does not always result from sacred events. What do the Taoists say? Chop wood, carry water, revealing that it is often the mundane which moves us toward enlightenment. The morning portion of the tour consisted of visiting beautiful natural settings and chatting with our guide. In the afternoon, we drove approximately fifteen miles down a dirt road to a site called Silver Falls. By the time we had arrived, the sky was dark and threatening, and it looked like it would start pouring any minute. However, we piled out of the van and proceeded to climb the mountain to the falls.

As we hiked, it began to rain and the ground became very slippery. It was already difficult climbing due to the thin air at 7,000 feet above sea level. The lightening and thunder were dramatic, and the combined thin air, threatening weather, and slippery ground forced many in the group to return to the van. However, I proceeded, slipping and sliding on the rocks around the waterfall as I pulled myself to the top under a sky of crackling bolts of lightening. I felt no fear! Usually, the mere thought of getting wet would have been enough to make me return to the van. I felt hesitant but enthused, impulsed to continue to the top.

I finally reached the summit and sat with our guide under a rock, letting the spray and sound from the waterfall wash over me. As I sat there, I could feel energy coursing through me. My fingers and fingertips were vibrating with energy, pulsing at a rate I had never experienced. I knew I had moved from segment five to six, and checked to confirm my intuition. Indeed, I was in segment six! I had gone on a guided tour, one I thought would pale in comparison to the sacred site I would visit at Mesa Verde, only to end up at a special sacred site in the middle of Pagosa Springs. I reminded myself that enlightenment can occur anywhere, even when playing the role of a tourist!

Elevating Peacemaker over Warrior Energy

I heard from the editor of the metaphysical magazine that was publishing my articles while I was on vacation. She had received the tape from our first channeling session and had listened to it, leaving a message at my office to inquire about her own desire for recoding. I told my office assistant to give her the telephone number where I was staying. The editor had gone through a difficult week as she and her husband, the owner of the publication, had determined there were pieces of their business to be terminated. They needed to move into another phase of their lives. It was a stressful time since their new phase was undefined. She called me in Pagosa Springs to ask if I could shed some insight.

I started linking for her to give her information about her soul contract in preparation for her reading of the formal request. Apparently, there were dark energies involved in a portion of her business, and her guides had initiated a house cleaning prior to her and her husband undergoing the recoding process. She and her husband were to honor their soul contract by informing others of recoding. Apparently, they were also incarnate members of the Nibiruan Council.

The editor asked how they had attracted dark energy when they were Light beings who had come to Earth to accomplish good. Nephras showed me how she and her husband had partnered for eons to fight the dark energies. They had often come up against the formidable Marduk on Nibiru when he had turned a reign of Light into that of darkness by dominating others through fear. Apparently, this couple had made a soul contract to battle the dark forces as a team. I saw them bringing dark forces upon themselves for the purpose of exposing them through conflict in lifetime after lifetime. However, their guides were telling me it was time for them to nullify their warrior contract and become peacemakers since it would interfere with their recoding contract.

The editor was relieved to hear this information because she was ready to do something different. I asked her to repeat the formal recoding request as I read it to her over the telephone. However, after she read the request, nothing happened. This was unusual for me because I always check to make sure that someone will receive approval prior to having them read the request. Yet, it was not being accepted. Then, I heard Nephras say that my new friend needed to formally request removal of her other soul contract from the Akashic records prior to her new one being approved. We were told that past attempts to bring higher consciousness to Earth had been unsuccessful due to the battle of Light against dark. This time, we were to proceed as Lightworkers without concerning ourselves about the dark forces. They were free to do what they needed to do, and we were to do our thing. There were to be no more conflicts or attempts to subjugate the dark by the Light. Actually, darkness is merely the absence of Light. This would ensure our success among those who opted for the Light, leaving the others to whichever path they chose. I finally understood why my guides were intent on moving us quickly toward segment nine so that greater consciousness existed on Earth prior to Marduk's return but, at the same time, doing nothing to stop Marduk from returning.

Dark forces and evil are illusions in the sense that Light dispels ignorance and intolerance which, in turn, counteracts the influence of the dark. Since we all originate from the Divine Creator, the source of all Light, how can darkness exist? I think darkness occurs as a natural part of the spiritual path to give us contrast so we can make enlightened choices. We put so much stock in darkness because of fear that we end up creating a dark and fearful state of being.

I gave the editor the additional words the guides gave me to nullify the other contract and saw her request for recoding finally receive approval. She asked when her husband could undergo the same process. I told her to fill him in on our conversation, work with him to remove his prior soul contract, and have him call me the next evening to read the request. I also told her to be specific when removing their former contract, specifying that she and her husband wished to remain together as a team, but in the new rather than the old contract. She told me she would begin supporting recoding by publishing my article on the *How Tos of Recoding*.

Acceptance versus Forgiveness

As expected, the owner of the publication called regarding his own recoding. Prior to him reading his request, I recommended we look at the barriers their publication had recently experienced. The latest issue had been in limbo for so long, experiencing one problem after the other which delayed its publication date. He said he had great difficulty with an article regarding forgiveness after he had written it. I linked telepathically with Nephras to gain insight for him.

Nephras explained that forgiveness was not part of Creator Truth because it was a value that assumed that someone had done something wrong. Right and wrong are an outcome of the polarity of this planet, and it breeds judgment which separates beings. Separation keeps love from growing, thereby weakening our state. Nephras said a more appropriate value than forgiveness is *growing the Light* because it allows everyone to exist and learn at their own rate. Acceptance was a value we must embody to successfully recode since past efforts at growing the Light have been thwarted by the dark forces. In this round, everyone will follow their calling, whether dark or Light, because acceptance is the key to success. This also ensured that fear would not be present since this was a negative frequency which would undermine a recoder's success. Hopefully, as the Lightworkers grew in strength, they would draw the dark toward the Light through example rather than coercement.

I remarked about how this information aligned with the information given to his wife the previous day regarding why they needed to disavow their warrior contract. We finished the session by having him read his formal recoding request. The pieces were now in place to align us, and I awaited more information.

An Unexpected DNA Fusion

While vacationing in southwest Colorado, my mother suggested we visit Lion Canyon which is a sacred Ute Indian pueblo near Mesa Verde. We had originally intended to visit Mesa Verde but were told that 700,000 tourists visit the site between June and August, and there is more irritation from crowds than peace from the sacred site. In fact, most of the sacred energy has apparently disseminated. We had heard of a tour company in Pagosa Springs that was authorized to bring visitors to Lion Canyon. Only 5,000 visitors were allowed to enter yearly because the Indians wished to preserve the integrity of the site yet allow non-Native Americans to learn about the importance of their culture. We opted for Lion Canyon over Mesa Verde and joined with a tour of 20 other visitors.

On the way, our guide told us about the mountain lions that roamed the canyon in the past, hence its name. Knowing that lions were closely aligned with the feline race from Sirius A, I asked Nephras about my connection to the canyon. It was really handy to have access to an angel who answered questions at a moment's notice. Prior to the indigenous Native Americans residing in Lion Canyon, Nephras said the entire four corners area was connected to the Sedona area where I had resided as a crystalline deity during the Sedona project. I knew this was not my first attempt at helping a third dimensional planet evolve to fifth dimension. The previous attempt was thwarted by Reptoids who came to take the knowledge residing in the Halls of Amenti, the power booster we need for this venture to be successful when the entire planet shifts. My immediate assumption, in hearing that mountain lions had been present, was that the Sirians were involved at some time. They had left the lions as guardians of their energy when they left the area. Now, hardly any mountain lions were left as civilization had enveloped the area.

As we neared Lion Canyon, I knew the day would bring unexpected experiences remembering my close ties to my Sirian heritage. What better space to work on me than in a location infused with energy from my lineage? My anticipation grew as we neared the canyon. A golden eagle arose from the brush and flew along in front of our van for several hundred yards. It was a beautiful sight to ride behind this magnificent creature who had a six-foot wing span. Our guide was delighted with the eagle's appearance saying there were few eagles left in the area and it was not the season for them to appear.

Lion Canyon was, indeed, a unique experience. The Ute Indians believed everything should be left as it was found, feeling every animate and inanimate object went through cycles, and that those cycles were not to be disrupted. The ruins were entered via the same narrow paths that the cliff dwellers used, accessed by long ladders that were hot to touch due to the midday sun. However, once in the canyon, it was cool and pleasant. The ruins had not been reconstructed, and fragments of bone, corn cobs, and pottery shards were available to touch as long as we replaced them. What a treat! Our Indian guide was the son of the Ute Reservation Manager. He was wonderful because he allowed us to experience the canyon without restrictions other than respecting the environment.

We began a one mile hike to Eagle's Nest, the highest cliff dwelling in the canyon. As we proceeded along the path, everyone became comfortable with their own pace so we no longer traveled in a group. I had been hiking the mountains all week and had adapted to the higher altitude so I was not short of breath like some of the others. However, about a half-a-mile into the hike, I experienced severe vertigo, broke out in a cold sweat, and momentarily blacked out. My mother who was sixty-five at the time didn't have any problems other than a little breathing difficulty. I knew this was the unexpected experience I had been awaiting. I asked Laramus what was happening, and he told me they had just fused my eighth strand of DNA. They chose to fuse it in the canyon where the energy was extremely favorable. Well, great, but I felt LOUSY! It took some convincing of my mother and the guide, who had caught up with us by now, that I was all right and could continue. I felt shaky but quickly revived and was able to enjoy the remainder of the tour.

Recoding beyond Kansas City

We went to Durango to do what all good tourists do when they visit the Southwest, spend money on Native American merchandise. We walked around Durango, browsing the shops and visiting the original hotels that have been renovated to replicas of the Old West. I did not find anything I cared to buy since most of it was jewelry and clothing, and I was more interested in spiritual relics. In the afternoon, we entered a small shop that contained the type of Native American crafts that I liked, beautiful animal sculptures, dream catchers and fetishes that looked individually crafted rather than mass produced. There was a dream catcher that immediately caught my eye, filled with crystals, animal skins and other natural elements. It was a very powerful piece.

I asked the woman in the shop the price of the dream catcher. It was about $100 more than I wanted to spend, but I told her I would be back if I wanted it. The dream catcher nagged me all afternoon. I knew it was something I had to purchase. I returned to the shop, and the owner laughed when she saw me. She said she had just finished telling another person who wanted the dream catcher that it was sold because she knew I would return for it. I was impressed that someone would turn down a sale for one that had not yet occurred. I thanked her for her faith in my interest despite my lack of commitment. We chatted briefly as I purchased the dream catcher, and she took the shipping information and gave me her business card. Her name was Donna. When I left she hugged me and said she hoped I would enjoy my purchase, a gesture I found strange since I did not know her at all. I knew the minute she hugged me that her energy was Nibiruan. I left in a rush to meet my parents so we could return to Pagosa Springs.

My guides nagged me all the way back to our resort. Call Donna, call Donna, call Donna. I could not get them out of my head! What was I going to tell her? Was I supposed to call a stranger and tell her she was Nibiruan? How would she react? This had never happened to me, and I did not know how I would handle it. But, I knew I would call her because I could not stand the noise in my head. So much for wanting to be able to channel: now I had more company at times than I wished!

I called Donna when we returned, told her I had just been in her shop, and asked her if she knew the terminology "starseed." She said she did. I asked her if she knew she was a starseed. She said she had guessed that she might be but did not know for sure. I informed her that she was from Nibiru and I had been told to contact her. I felt bizarre imparting this information, but so far, so good. I asked her if she knew about recoding, and she said she had read some books about it. Thank goodness, our guides prepare us for what is coming! I told her about my involvement and promised to send her my writings about the levels of recoding and the tape from our first channeled session. I told her to call me if she was interested after she had reviewed it.

Donna did call several months later, and said she was interested in staying in touch. I worked with her spiritually, conducting several telepathic sessions as well as a soul-clearing on her following my workshop in Colorado where I learned how to do so. Through Donna I also met, via telephone, a group of individuals who lived in Durango and were interested in soul clearings and recoding. I have worked with all of them, helping to establish a network of recoders in another part of the country. What an odd experience from the day I met Donna until later when I had worked with many of her friends! I never knew when I was going to meet another starseed or what would clue me into their origins. However, I had learned that I needed to trust the situation and listen to my guides. They were not going to embarrass me, and they ensured that others were prepared for the subject. Again, my issue with faith in my guides arose. I promised myself I would spend more time believing them and less time analyzing.

Creator Empowerment

N called to tell me she had moved to segment seven, "Removing Illusions," and had some preliminary information about it. I eagerly asked her to share what she knew and she described it as follows: one will feel the presence of a light at the end of a long dark tunnel which symbolizes one's mission, and one must traverse the tunnel to reach the light. One's personal power is the torch that lights the way toward one's mission. Every time you hit an obstacle, shine the torch on it, thereby exposing the illusion that has been created.

Later, I asked my guides to add more information on segment seven since the gateway had been opened. They told me that in segment six, one learned how to be a maker. Now, in segment seven, one has become a creator. Creators manifest any reality they choose since they realize they create their world by placing attention on what they want. Whenever they hit a perceived obstacle, they shift to a more beneficial reality as "creator" by focusing on what they want rather than what they don't want. Unfortunately, one also acts as a creator when hitting a perceived obstacle and choosing to identify it as a problem. As the creator, one can create a different reality so it was important to remember to choose the path of positive rather than negative perception.

Makers versus Doers

I remained confused about segment seven. I finally asked Laramus for clarification. He said that first I needed to understand segment six. Laramus said segment six was tricky since we were no longer relying on our guides to generate manifestation energy for us. We were finally going solo. This was different for most of us since we had been using creative visualization as a means to manifestation. We had been taught to state affirmations in present tense and ask our guides to use their energy to support our requests. However, after reaching segment six, my guides had been gently (and sometimes not so gently) reprimanding me to use my own energy.

For example, if I desired financial abundance, I was to experience the feeling of having the abundance by generating the emotions that were part of that situation. I needed to *feel* rich to be rich. If I wished to have my life mate Asalaine join me, I should view myself as a radio that broadcasts its frequency and sends messages to my future partner, feeling my energy projecting in his direction. Also, I was to support that broadcast signal with the feelings I would have while with my life mate, experiencing the emotions of sharing time or making love with a significant other. The more intense the feelings I generated, the faster I would be able to create what I desired.

For those who have read _The Tales of Alvin Maker_, it reminded me of Alvin because now we are the makers rather than passive receivers of what our guides do for us. I had been stating my affirmations, expecting my guides to intercede on my behalf rather than using my own energy. I was becoming what my guides had always been to me! I understood that I could always ask my guides to support my efforts as they could not intercede unless I requested. However, my guides now played the role of assister rather than maker or doer since that was my role. Once we understand that premise, we are ready for segment seven because we are no longer thrown by unsavory events. We understand we have the choice to include them in our experience or edit them for a more joyful experience. We are totally responsible for every experience we have.

I was excited to receive this information because I knew I was capable of manifesting but was unsure how to unleash the energy. I decided to experience my manifestations emotionally, which is essentially energetically, by focusing on manifesting my life partner.

Soul Groups and Lineage

I asked N to channel the male energy that was responsible for my soul inception. This session resulted in a wealth of information, including clarity regarding my lineage. Based on my being a mutable, which was a soul designed to communicate with all beings in the galaxy, I surmised that I had to have reptilian lineage. How else could I have intercepted the information regarding the reptilian attack on the crystal city of Sedona? Yet, my previous channeled sessions had only surfaced the Sirius A and B ancestry as well as the Nibiruan/Pleiadian ancestry. The being that emerged called himself Natara and was able to answer my questions.

Natara said he had originated from Sirius A, the feline people who called themselves the Elders or Creators because they were responsible for seeding other planets with new civilizations as they had already completed their work in their own universe. I asked if he had any association with Lion Canyon, the location I had recently visited on the Ute Indian reservation, and he said that was an energy point for Sirius A. He also confirmed they had worked on my energy while I was in the canyon because I was within close range of their frequency.

I asked Natara how he could be from Sirius A when I had been told that my originator was a galactic human from Sirius B. He said that I was fostered by my Sirius B father, named Gunnaia, but I was actually a fragment from Natara. He explained that the Divine Creator had desired to experience more aspects of life so it released fragments of itself to seek other types of dimensions. Those fragments then fragmented themselves to increase their repertoire of dimensional experiences. I asked how many fragments existed, and Natara said they were too numerous to count. I also asked how the fragments returned to the Divine Creator. Natara said the fragmented souls returned through the individual Divine plans that were created for each of them.

Natara said he had been able to mate as a eleventh dimensional being with a sixth dimensional female to produce my soul because he worked through the sixth dimension which was the level of creation for all civilizations. Once my soul was created via the two fragments, it was fostered by my Sirius B father Gunnaia. Obviously, by now I was confused, and Natara acknowledged my confusion by saying that it was a complex process and beyond my current understanding. I asked how the reptilian piece of my heritage had been created, and he said that my physical vehicle had been seeded with the DNA. Natara said the ancestry of souls becomes even more complex because, for example, my son Drew's soul originated from Natara. However, Natara did not work with a female energy to create Drew. Drew originated entirely from Natara. I asked Natara if love was an aspect of these soul "matings" or were they more aligned with test tube pregnancies. He assured me love was always an important element, since without love, the new soul's frequency would not be complete.

Natara said Asalaine, who would come to Earth to be my life partner, was part of my soul group. I asked him to define soul group, and he said it was a group of beings who resonated at the same frequency and made a commitment to walk the same evolutionary path. Soul groupings could be large, like the Nibiruan civilization, or they could be smaller subgroups existing within subgroups of the larger soul group. Natara said all of my soul group was not on Earth at this time, but we would eventually be brought together since we were all working on the same plan. I asked him, what plan? He told me I would remember eventually, but they had purposely blocked my memory since the pull would be very great for me to leave Earth if I remembered. He said it was sufficient for me to know they were supporting me in my soul contract and would allow me to remember when they were assured I would not try to return home. Once we vibrate at a higher frequency, we have the ability to walk through a dimensional portalway that takes us back to our original, higher frequency state of being.

I asked Natara if completion of my contract would "promote" me to a higher dimension. I was unsure what else to call it, but Natara understood my question. He said I would move well beyond my current dimension when I left Earth, but the level depended on my ability to serve. Now, this was tricky because I was still unsure exactly what I was supposed to be serving. Natara must have heard my thoughts because he said I honored him through my work since I was a fragment of him, and they did not choose the members of their mission lightly. He said it was joyful to speak with me, and, now, Anu desired to speak with me.

Anu returned and said I would be awed if I realized the amount of protection around me at this time. He said if I could "see," I would realize how much assistance I was receiving for this mission. Anu said good bye, and N returned.

More Walk-ins among Us

My roommate Julia discovered she was a walk-in, a shocking experience for her since we determined it was recent. It is quite nerve-wracking to discover that the person you thought you were is not the person you are at all. This happened to me when I discovered I was a walk-in. I started thinking about the character traits I had prior to walking in and the ones that surfaced afterwards. There were clear differences that emerged once I thought about it. Yet, since we carry the memories of the original soul who resided in the body, it is difficult to separate who we are versus the original soul. It is the ultimate identity crisis!

Here is what happened with Julia. I had been taking a correspondence course on soul-clearing and de-possession from my teacher, Venessa. In fact, I was scheduled to attend an apprenticeship weekend in early August which I anticipated with excitement. I had been listening to Venessa's audio tapes and reading the materials that accompanied the course. I was at the point where my homework required reading the Akashic records for five people and sending the information to Venessa for review.

I asked Julia if I could read her soul records for practice. To read the soul records, one must first determine that the correct soul information is being tapped. Therefore, one of the first questions is whether this person is mono-souled, and has a single soul existing in their body for this lifetime. Since Julia had already had a soul-clearing several years before by one of Venessa's students, she expected a "yes" to this question. However, I did not receive a "yes." We were perplexed, knowing she had a single soul residing in her body or we would have noticed a different type of behavior. I decided to break the question apart and restate my query to try to clarify the information. I received a "yes" to the mono-soul portion and a "no" to the portion regarding a single soul residing in the body for this lifetime. Immediately, I realized Julia was a walk-in, asked for confirmation, and received it.

Julia stared at me, protesting the information. I quickly received additional information that she had walked-in within the past year and asked her if anything unusual had happened last August or September. Suddenly, she recalled a road trip she made to MacPherson, Kansas during which time she kept fearfully repeating to herself as she drove that she was "going to die." When she arrived in MacPherson, she blacked out for thirty minutes during which time her friends tried to revive her. This is typical of a certain genre of walk-ins who make the swap during one propitious moment. It is important to understand that these are not possessions, and that a walk-in always has a previous contract with the soul residing in the physical body. Yes, I confirmed, this was the walk-in event.

Julia was distressed, trying to understand what part of her was Julia and what part was someone else. I decided to link for her to provide some answers. We discovered that Julia was born in a soul grouping of four, much like quadruplets. One soul was her husband whom she was currently divorcing, one was the original Julia, one was the current walk-in, and one was the man she would meet who would become her life mate. The original Julia had a contract to bring her to the point of emergence, surviving an emotionally abusive childhood and a difficult marriage, but not breaking ties with family or husband. The walk-in was to enter Julia's physical body without remembering who she was and bring Julia to freedom. Again, this is typical of walk-ins who must clear the karma and achieve the soul contract of the original soul as well as their own soul contract. Most walk-ins are highly evolved souls since this requires doing double duty during a single incarnation. I was not surprised that Julia received this information on July 22, the very day she filed for divorce, since that act completed her contract as well as fulfilled the walk-in's contract.

I asked for the walk-in's name and was told it was Pelauria. We proceeded to complete the Akashic record reading to determine if Pelauria's archangel realms and soul societies were different from Julia's. Not surprisingly, they were entirely different. Pelauria carried a different soul imprint with different experiences than Julia. Pelauria would need to get to know herself.

Carrying Polarity Baggage

Again, I attempted to proceed on my soul-clearing homework assignment, this time asking my son, Drew, to work with me. I did not know what to expect after the previous night's experience, but decided it was important to practice. Again, I experienced a glitch as Drew held a compassionate connection with a dark energy from a previous lifetime despite the fact that he was a Light being. This dark energy had the capacity to draw him toward a dark vortex and needed to be removed.

Now that I have conducted numerous soul-clearings, I have a greater scope of understanding regarding the amount of dark alliances and karmic connections we carry in our soul imprint. As fragments of the Divine Creator, we left the Source to increase our experiences while simultaneously expanding the experiences of the Creator. This meant we lived both Light and dark existences in order to experience polarity which gave us a point of context for our origination point. However, in the process of experiencing darkness, we sometimes created bonds that were never broken. These follow us into our Light lifetimes and sometimes create problems we cannot resolve on a conscious level. Had Drew continued to live with a link to the dark side, it might have negatively impacted his life through unconscious influences that eventually surfaced into consciousness. I proceeded with the request to clear his association, happy to release him from a dark presence, but not realizing the implications this type of work had for me.

Experiencing a Psychic Attack

Clearing my son Drew from the compassionate connection he held with some dark beings led me to some of my own life lessons. The night after clearing him, I dreamed about being pursued by creatures that looked like a hybrid lobster, scorpion, and snake. They were evil, mean and relentless. I was fighting and kicking them away with all my might, feeling their weight to be about the size of a small dog as I contacted my foot with their body. All of a sudden, I awakened, realizing that this was not a dream and these horrible creatures were trying to enter me through my feet. My heart was beating furiously as I tried to shift white light through my body, but I encountered two problems. First of all, I was too depleted of energy due to the negative energy that had entered me. Second of all, my arms and legs were paralyzed and my head felt crushed by a tremendous pressure. Part of my astral body was still away, being altered by the nightly work conducted by my genetic engineer, Laramus. I had partially returned to protect myself, but Laramus did not have time to return all of me.

I was a mixture of me and my copy who managed my physical body while my astral body was gone during my sleep time. Although my copy was a different energy, it resided in my physical body so it ran off my programs. The copy's responsibility was to energetically run my body. However, the programs used to run my body were still my own. Therefore, the copy made me *feel* different. It did not have the same response mechanism that I had since it had never experienced fear—at least until that very moment.

I immediately called out to Jesus, Archangel Michael, and Archangel Gabriel to help me. I asked them to push the negative beings out of my body. I felt them being removed as enough of my energy returned to enable me to assist in the extraction. I still felt lousy, but I was also angry. How could my guides leave me unprotected? I trusted them! Obviously, my copy did not have the same protection instincts. I told my guides they could not work on my astral body if they did not adequately protect me. They apologized for the event, but somehow I felt there was a lesson in it for me in regard to holding my power. There was also a lesson for them. They realized they needed to train copies to combat negative energy recoders might attract rather than leaving the recoder exposed and vulnerable while sleeping.

The next morning, I called Bobbie for advice. She said she did not know I was "traveling through the universe" doing soul clearings. She said I should know better than to tackle dark forces without protecting myself while sleeping, which is energetically a vulnerable time. Well, I did not know this but vowed to never expose myself again. She told me to ask my healing guides to assist me by running light energy through my body continuously for several days as she saw traces of the negative energy still with me. She had also seen these entities during her own psychic work and confirmed they were vicious. I immediately solicited help from my guides, and I felt my feet tingle all day as the energy they sent coursed through me. My head ached, but it was slowly improving.

Protecting ourselves energetically is an important lesson for those recoding. First of all, we attract dark energy that is looking for a competitive playing field. Battling strong Light beings is fun stuff! Secondly, we must accept our power. I have always been squeamish about the dark side and have said that it is not my job to get involved with them. However, if I were to own my power entirely, I must overcome any fears I have, whether physical or energetic. I asked my protector guides how to ensure I did not encounter any other dark energies while sleeping? They explained how to create an impenetrable force field around my body before falling asleep by envisioning two opposing fields of white light that met in a spherical shape around my body. The point of resistance between the white light inside the sphere versus the white light outside the sphere would act as a barrier to oppose any dark energies. I was grateful to receive a method for protecting my vulnerability, although I did not envision sleep would come easy for a while. I had felt jumpy all day, and my scalp kept tingling like invisible energies were brushing against it. I kept remembering the vicious appearance of the scorpions and the thud of my feet against their bodies, shuddering at their frightening demeanor.

Exposing Illusion

I was in segment seven, arriving after making the decision that the scorpion creatures could not harm me. Remember, segment seven is about creating one's own reality by exposing illusions, and the scorpions were my first seemingly real illusion of dark energy. I had a very difficult time falling asleep after my run-in with these creatures. I was in a hotel in Charlotte, North Carolina following some research for an east coast client. As bedtime came and went, I realized how afraid I was of falling asleep, despite the fact that I had been taught how to protect myself. I toyed with the idea of calling Bobbie for emotional support, but it was past midnight, and I did not wish to awaken her. I decided to brave it alone. I reminded myself I really was not alone since my guides were always there to assist me. I asked them to remain with me during the night and protect me.

Then, it occurred to me that I needed to get over it. If I was to recover my power, I could not be whimpering in my bed, afraid to shut my eyes due to dark energy. All energy has a pattern, and that pattern can be manipulated according to how we manifest our own energy. I could not allow a dark force to cower me by thinking it was more powerful than I was. I had to believe in my ability to protect myself without the assistance of my friends. I decided to build my protection and fall asleep, confident that the scorpions or other such entities could not reach me if I chose not to allow them near me. Besides, Laramus was clamoring for me to fall asleep so they could complete their work that had been surprisingly abbreviated from the previous night. Although I was unsure how soundly I slept, I placed protection around me and willed myself to sleep even though it only lasted four hours.

When I awoke, my entire body was vibrating. My feet were buzzing, and I was experiencing intense spasms throughout my body. At first, I was frustrated, asking my guides why they had not done a better job of correcting the previous night's work and putting me back together. I heard them tell me they had made repairs, and I realized I was vibrating because my energy was getting lighter. I was in segment seven. I had shined my Light or power on the scorpion illusion of danger and fear, and I had squelched it.

I congratulated myself, wondering if I had really conquered segment six since I had yet to manifest my financial winnings or Asalaine. But, I also was beginning to understand that attaining the various segments is a cumulative effect that does not necessarily complete the prior segments. In other words, I was in segment seven to begin working on aspects of it, even though I had not completely graduated from segment six. I was still learning how to manifest immediately and accurately. Now, I could work on manifesting and creating simultaneously.

I lay in bed, feeling the round of spasms that accompanied my new vibratory rate. I asked my guides how long the spasms would last, and they said "forever." They explained the spasms were a result of a lightening of the third dimensional energy and the spasms could not be controlled. I was disappointed because the spasms were very annoying. Then, I realized they were continuing to teach me the creator aspect of segment seven. Nothing is forever if we do not choose for it to be that way. I told them I refused to accept eternity as an answer, and they better return to the drawing board to rework the solution since I expected them to find one. They laughed at me, delighted I was recognizing and appreciating my new role. The spasms stopped within the day.

Moving into Segment Eight

N came over to check the taping equipment for our second channeling session on Sunday. She had just moved to segment eight which was called "Releasing Fear," and I asked her to share what she knew thus far. N said segment eight was the time when most of the work was conducted on how our brain processes information. By segment eight, most of our DNA strands are fused and activated in our astral body. At this point, I had eight fused, and I was in segment seven. By reprogramming the way our brain communicates information, the DNA is able to carry the additional information throughout the physical body. The entire firings or synapses of energy sequences operate differently once this work is completed.

N said life had become incredibly easy for her. She was finding opportunities at every turn, both business and personal. Whereas segment six was a time to take back power and begin feeling the importance of our role as a maker, segment eight provided the juice to immediately manifest. This was when we reap the rewards of learning how to harness our own energy rather than being on automatic pilot while our guides bring manifestation energy. When we recognize the potential of our own will, we can do wonders.

Dealing with Reluctant Caretakers

Response to recoding continues to be interesting. Some gravitate to it with a passion, grabbing any opportunity to learn more about it. Others are casually interested in it but do not see it as a key life event. Still others attempt to proceed through the segments but waiver in determination as the challenges mount. Personally, I was fully committed to the process and continued to take people through their soul contracts and the approval. However, I also empathized with those who found the path stressful.

There were days when I wanted to return to a less tumultuous life, one filled with fewer events. And, there were times when I felt crazy since I spent more time having conversations with beings who were not physically present than those in my everyday life. Then I remembered the mundane existence facing me if I did not recode, one filled with fear, anger and routines rather than joy, love and flexibility. I also reminded myself that even a mundane life was at risk as we moved into our upcoming dimensional shift.

I had been moving through segment seven, and it has been the most difficult level thus far. Segment seven is the one where we must use our growing power to light the way through the illusion of obstacles. Obviously, my fight with the dark energies who "slimed" me was part of that course. However, there had been other setbacks with which I had less success. For example, the editor I had assisted in recoding called to inform me that she and her husband had decided to sell their publication. Unfortunately, all of the articles I had expected to read in the upcoming issue were on hold until a buyer appeared. I was extremely deflated by this news since I had worked so hard to meet the short deadlines to be well-represented in the publication. I had not quite accomplished the knack of living without resistance, shifting attention from what I did not want to what I desired when a block occurred.

Segment seven was also a time when our astral bodies are taken away for longer periods. It was necessary for our genetic engineers to work on us for two or three days at a time, because they could not do everything during short stints when we were asleep. They replaced us with a copy, or "caretaker" as they preferred to be called. These caretakers certainly had their shortcomings as I eventually discovered. First of all, they needed to be trained or they would do whatever they pleased. This could mean they would do nothing when I was faced with an important work deadline, or it could mean they would binge on sweets after I had spent the last two years weaning myself from sugar. Apparently, caretakers are young souls who have never lived in the third dimension. They are coming here for the experience as part of their own soul growth. That is why my copy was unfamiliar with a fear response when it was warranted by the scorpions. Additionally, caretakers have no emotions since they are not vested in anything on this plane. They act rather spacey which was noticeable by those who knew me well.

During our second public channeling session, Joysia spoke at length about the frustration we felt with our caretakers, telling us we needed to train them or they would act more like a "house sitter" than a "housekeeper." He said the first thing we should do is protect ourselves by a net of gold light that we should zip around ourselves in the morning and at night. This was in response to my experience of having the bad guys invade my caretaker during the night. Caretakers have never experienced negative energy and did not know how to handle it. By protecting my physical body, I used my awareness of negativity to make a protective field. The caretaker could establish the same pattern because of the template I had created. It was also important to grant permission to caretakers to read our records since no one was permitted into our records unless we allowed it. By having access to our records, we had greater likelihood of the caretaker reacting similarly to ourselves in situations. Joysia also recommended that we leave a list of instructions for our caretaker so tasks were completed while we were "away."

Kryona was my caretaker, and I was trying to assist in her training. After experiencing an emotionless weekend filled with lethargy due to my caretaker's presence, I decided to have a talk with her. Besides, all of my friends were asking me what was wrong with me. They knew I was not acting like "myself." I realized Kryona was doing the best she could, but it was difficult to forego the joyful feelings I had come to expect as a part of the lightening of my energy. I told her she needed to try to experience joy on a daily basis while here, and she needed to be productive. I also told her I had worked too hard to give up sugar, and she needed to curb her sweet tooth. She agreed to cooperate. In fact, she wanted to occupy me full-time while I was doing my market research work. She felt that she could free me up for more astral body work in order to hasten recoding. However, I did not entirely trust her with my client base.

I struck the following agreement. I would allow her to occupy seventy-five percent of me and write a research report. If the report was good, I would allow her to maintain her seventy-five percent during work days over the next few weeks to complete my astral body work. However, I wanted to be fully present during weekends. Since recoding, I had discovered a joy for life I did not wish to give up during my free time. I also asked Kryona to feel free to do anything better than I did it if she felt more capable and to eliminate any bad habits. I figured I might as well maximize someone else's talents. Unfortunately, I could not ask her to be more energetic. The caretakers seem to have low energy levels, perhaps because they are unfamiliar with the density of this planet. I was taking naps during the work day, something unheard of in my typically high energy lifestyle. However, I was going to have to accept the low energy level since there was nothing she could do until she acclimated to the density.

Kryona wrote an adequate report for me although there were a few details missing. All things considered, she was a sufficient replacement. I left town with seventy-five percent of her and twenty-five percent of me in my body, deciding I would like to be partially present to ensure I did not lose any clients. I was starting to feel like Sibyl because I was carrying on multiple conversations inside my brain, some with my guides, some with my genetic engineer, and now some with Kryona. I was glad there was no one intimate in my life (Asalaine had not yet arrived) because I could not handle one more point of contact. I was also thankful my son was at camp for three weeks since it was hard to convince those who knew me that everything was fine. I checked to see how long I would need to endure my multiple personality and was told the work would be completed during two three-day sessions. This was not a bad time commitment when considering the positive outcome from recoding.

Summary: July Recoding Lessons

- All beings in the universe are tied in some way to Earth, as their evolutionary return back to the Divine Creator is hampered until third dimensional beings raise their frequency level.

- Recoding is a key part of reducing the effect Marduk will have on this planet if he returns. This event may occur in our lifetime. The higher the energy frequency that exists on the planet when Marduk returns to reestablish fear as a primary motivator, the greater the likelihood of Earth's success at moving into the enlightened 2,000 year cycle predicted by ancient civilizations.

- Fear can undo what recoding has done during one's progression through the segments. It is very important to remember that fear is an illusion established on Earth to restrict third dimensional beings from realizing their true potential.

- Recoders will need to rid themselves of the majority of their negative emotional energy, mostly in the form of anger or fear from this life as well as past lives. In fact, there may be physical discomfort during some of the work depending on how much negativity has been cleared prior to beginning recoding. It is recommended that one use liver cleanses for removing and keeping anger from the body since the majority of anger is stored in the liver.

- During recoding, it is beneficial to eat as little meat, fowl, and fish protein as possible in support of cleansing the physical body for recoding since meat and fowl is filled with antibiotics and hormones and fish is filled with ocean pollutants. These chemicals interfere with the cleansing of the body, creating a continual dosage of negativity in need of elimination.

- It is important to release those whom you feel anger toward rather than forgiving them. Forgiveness is not part of Divine Truth because it sits in judgment, assuming someone has done something wrong while the other has done something right.

- Segment three is when the clairaudient channel begins to open. If one resists what they begin to hear, there may be pain in either the right or left ear depending on which ear is used for telepathic communications. It is important to avoid heavy electrical environments where the electrical current flows in a random state because it causes static in the telepathic channel and may result in nausea or achiness.
- Not everyone will see higher frequency entities with the naked eye in the physical plane after clearing and completing segment four. For some, these entities will be seen psychically through the third eye.
- During segments five through seven, caretakers will be left on Earth for longer intervals since the genetic engineers will need one's astral body for longer periods. It is important to make a "to do" list for a caretaker in order for them to be productive during a recoder's absence. Caretakers are young souls who wish to experience life in the third dimension. Caretakers have no heart connection to one's life since they are copies, and they will be more neutral and passive. It is helpful to give them permission to review one's records, so that they have a frame of reference upon which to draw.
- Segment five, "Integration," is exciting because one can begin to transfer to other dimensions, simultaneously experiencing multiple lifetimes. During this time, bleedthrough might occur which may come through in the form of a dream or flashes of deja vu.
- By the sixth segment, "Owning Your Power," the guides are ninety-eight percent positive that a recoder is firmly on his or her path. The recoder begins to receive greater levels of power. It is the segment that enables the mind to manifest what it desires quickly. This is a time to maintain positive thoughts since negative thoughts can also manifest. One's guides now play the role of assister rather than maker or doer since that is now the recoder's role. However, at this point, less than half of the twelve strands of DNA are

connected which limits one's ability to move to complete manifestation.

- The seventh segment is called "Removing Illusions" and is a release of multiple layers of density, peeling each layer back just as layers can be peeled from an onion. This is when one's density begins to truly lighten. In segment six, one learned how to be a *Maker*. Now, in segment seven, one has become a *Creator*. Segment seven is difficult because one moves from the realization of oneself as the power source in segment six to running the gauntlet to test that power.

- By segment seven, one will probably begin to attract the attention of dark forces. They are mesmerized by this type of Light energy since it is ultra-powerful and would be a tremendous boost to their own energy. If one is not already performing a daily protection routine, now is the time to start.

- Segment eight is called "Releasing Fear," and it is the time when most of the work is conducted on the way the brain processes. By segment eight, most of one's DNA strands are fused and activated in the energy body. The firings or synapses of energy sequences throughout the brain will now operate differently. By reprogramming the brain, the DNA is able to carry the additional information throughout the physical body.

Chapter Five

August, 1996

Feeling Bad during Recoding

Segment seven, "Removing Illusions," was the most difficult segment I encountered; difficult because I did not know if it was me or my caretaker reacting to situations. I could not discern between my responses and Kryona's, so I was unsure if I was solely responsible for my actions or if another soul was involved. I had felt dull and uninvolved since Kryona began caretaking my body, a reaction to my caretaker's lack of emotion for a life she has not lived. I was present, yet not present.

I had set the condition for Kryona to leave for the weekend so I could enjoy myself. I missed the positive, light feeling that I naturally brought to myself when I was whole; a feeling that Kryona was incapable of manifesting. However, I discovered upon returning from my business trip on Friday that I had to fall asleep for my caretaker to exit, and Kryona had decided she did not wish to leave. She liked Earth and wanted more experiences! I had been invited to a gallery exhibition and just had time to shower and change between the arrival of my flight and meeting with my friends. There was no time to fall asleep to swap out Kryona. I went to the exhibit in our seventy-five/twenty-five percent configuration, but typical of Kryona's energy, felt disoriented and had difficulty connecting with others.

On my way home, I rented a video because I knew I could not sleep, partially because Kryona wanted more time, and partially because I had developed a pounding headache that made it

difficult to relax. When I returned home, I took a hot bath, hoping
it would relax me enough to sleep, but I emerged equally agitated.
I decided to watch a video which ended at 1:20 a.m. I was still
wide awake and, by now, my headache had traveled down the
side of my neck. I seldom take any medication, but I knew Kryona
would not leave until I slept so I took two extra strength aspirins,
hoping when I awoke in the morning I would feel like myself
again. I finally fell asleep around 2 a.m.

When I awoke, I felt very depressed and had no energy. I
thought perhaps Kryona was still present, but I checked and
determined she was not. How could I be back and feel so low?
Then, I realized that a major percentage of me had been with my
"real" family for almost a week. Despite the fact that my memory
was blocked, I was profoundly homesick. I could not pull myself
out of it. I spent most of the day floating in the pool, trying to
orient myself. I was also depressed about the recoding process
because I knew I had ten strands of my DNA fused and wondered
when I would start feeling "different." My friends did not relate to
this attitude because they knew how psychic I had become over
the last few months and how quickly I manifested things. I still felt
less evolved than I expected I would at this point. How could an
emerging fifth dimensional being feel this bad?

On Sunday morning, I spoke to N as the depression had not
cleared and I was growing increasingly concerned. She empathized
with me, reflecting how difficult segment seven had been for her
as well. Segment seven exposes all remaining fears residing in our
system, forcing us to address and remove them. When we come
to Earth, we have karmic implants that are placed in our body to
assist us in learning the lessons we have come to experience. Some
of those implants actually enhance the level of certain fears to
supplement our learning experience. I guess we graduate magna
cum laude if we overcome the fears that have been intentionally
magnified. This seems unfair to me from the third dimensional
perspective, but I guess as a non-physical being we enthusiastically
seek challenges in order to ensure we completely overcome them.

During segment seven, we are forced to face the deepest
fears we came here to conquer. Then, they can be removed. My

remaining fear was a fear of being controlled by others through mental or physical possession. Well, they had certainly tested me on that one the week before with that nasty scorpion energy. I had also been tested via my caretaker who did not want to leave. N advised me to "let go" of everything in order to bring segment seven to closure. I did not know what "letting go" meant in this context so I was unsure how to proceed. In the meantime, I repatterned any negative energies as they arose, staying as clear as possible during this time. Hopefully, I would grasp the meaning of "letting go" soon since I disliked living in a negative state.

Visiting the Halls of Amenti

Jeanne and Mark, the couple I had assisted in requesting recoding, came over to help me visit the Halls of Amenti. I had two reasons for asking their help. Number one, Jeanne uses a process called "flowing" where she is able to take people on guided visits to other dimensions. I had asked her if she would be willing to take me to the Halls of Amenti to find out how I was involved. Secondly, Mark had been part of the web, or grid energy, that was established as part of the Halls of Amenti which will be used as a booster to move the Earth from third to fifth dimensional consciousness. This was the reason he had entered his lifetime in "group think" regardless of the unpleasantness of the situation. He needed to understand how to interrelate energy to create the foundation for moving individual energies from one dimension to another, especially energies as seemingly separate as third dimensional beings. I had already shared with them the information I had received about the Halls of Amenti regarding Mark's involvement and my own. They were as interested as I was in exploring this sacred realm.

When they arrived, Jeanne asked if I had permission to visit the Halls of Amenti energetically, knowing that I was not currently allowed in Sedona due to the codes I carried in my body which could open the portal to Amenti in the future. I told her I had already verified we were able to visit, learning that traveling astrally would not interfere with the energy there. Jeanne prepared us for flowing, having the three of us lie down and hold hands with Mark lying in the middle. She verbally moved us into a meditative state, proceeding to take us to the entry of the crystal city, the one I had told her I frequently saw hovering over the current site of Sedona. We proceeded over a bridge that spanned a beautiful crystal waterway that meandered throughout the city, filled with tropical fish and careening waterfalls. We saw how the tropical fish of today contained the same crystal coloring of that spectacular city. We also saw how the water derived energy from the crystal nodes that protruded from the main temple, and how people used the water for drinking and bathing as a source of good health and spiritual nourishment.

Jeanne helped us proceed through the temple to the area that would take us to the Halls of Amenti. We saw a pyramid topped by a floating sphere to the left of the temple, with access through a "floating" walkway. I gasped when the pyramid and sphere appeared because it was my new InterLink company logo. I was stunned to see the past lifetime meaning those shapes had to me beyond their sacred geometric configuration. We entered the pyramid and descended a stairway behind the podium that led to the portal of Amenti. At that point, we diverged. Jeanne said, "we are moving down a long, narrow hallway and we are entering the appropriate door which is the second door on the right." I interrupted her to say, "I will proceed down the hallway as well, but I will enter the room that is straight ahead." Mark entered the room on the left.

Her room was filled with stone tablets in iridescent drawers that resembled a safety deposit vault. My room was filled with white Light which transitioned me into the actual Halls of Amenti. Initially, we were perplexed by our different locations since we had been traveling together until this divergence. Later, we realized we each had played different roles in the crystal city which is why we went to our respective positions. Jeanne had been the keeper of the codes that were inscribed on the crystal tablets. Mark was a crystalline deity in the Halls of Amenti. I was the keystone within the Halls. Jeanne said they would follow me to my destination. I entered an oval room that was encircled by thirty-two crystalline deities, Mark being one of them. Each one wore the Egyptian headdress I had seen on busts of Nefertiti or Tutenkamen. There was a frieze of different scenes around the top of the room that integrated the energy of the various deities. I sat in the middle and began communicating with all of them. Jeanne asked for those of the thirty-two we knew as Earth incarnates to please step forward. Mark was one of them and there were four others who we could not seem to recognize. Mark was told that it was his job to identify those of the thirty-two beings who were incarnate and bring them to awareness of their dual existence, one part on Earth and the other part in the Halls of Amenti.

Jeanne brought us out of the Halls of Amenti and out of our trance. The experience was awesome, more so because the three of us were connected by our work in the crystal city yet unaware when we initially met. In fact, Jeanne and Mark had met the prior year at a singles function at Unity Church. The sequence of activities that had brought us together in incarnate form was astounding. Additionally, I had never traveled someplace astrally with others who could see the same scenes. It was very powerful to experience the richness of my own visuals as well as having someone else simultaneously describe it.

The Nuances of Segments Seven, Eight, and Nine

I took a long walk in the woods to commune with my guides away from the interference of electrical currents. I told my guides that my motto for segment seven was "seven sucks." Yet, despite my difficulties, I had ten strands of DNA aligned and fused into my astral body. I wondered why the presence of so many strands did not ease my progress through segment seven. I was still fuzzy about the differences between the manifestation aspects of those two segments and the purpose for the stress imposed by segment seven.

I realized that segment six, "Owning Your Power," was a time to "take on" our newly acquired powers. It was a transition from the "help poor little me" requests made to our spirit guides, to the recognition that we owned our own power and needed to begin using our manifestation capacities. However, at this point, the unconnected strands of DNA limited our ability to move energy from a non-physical to a physical state. Segment seven, "Removing Illusions," required us to test our new skills acquired by the receipt of the existing DNA strands. We were thrown obstacle after obstacle which were associated with any past fears lingering in our consciousness. We had the choice of accepting or rejecting the obstacle based on the fact that any obstacle lying in our path is an illusion. If we have learned our lesson well in segment six, we understand that we manifest our own destiny and can create any reality we choose. Segment seven is tough because we move from the realization of ourselves as the power source to a running of the gauntlet. It is as if everything we believed in segment six is being put to the test. Yet, for me, segment seven was completed when I threw up my hands and said, "Enough of this, I don't choose to live this way." It was the "letting go."

Finally, we enter segment eight, "Releasing Fear," which clears the illusions from segment seven and allows us to completely manifest as long as we use our own power to create the manifestation and not the power of our guides. By the end of segment eight, all of our implants relating to fear are released. These implants were placed in our bodies prior to entering Earth in order to allow us to maximize our experience of the third dimension which holds fear as a prevalent emotion. Additionally, the fear implants protect us in this dimension as fear is not in the emotional repertoire of beings from higher dimensions. This is probably because death has no meaning to them since they can recall past, present, and future simultaneously. Finally, at segment eight we have ten strands of DNA to assist us in addition to our recognition that we are the ultimate commanders of our path. We recognize that the elimination of fear defuses our magnetism to violence, and that we are protected by our own ability to manifest clear emotion. We are closer to Spirit than we have ever been while living on Earth in a third dimensional reality.

I had assumed the additional strands of DNA would have provided me with more abilities by this time. Anu had predicted that some people might feel they could walk on water like Christ by the time they had cleared the nine segments. He had explained that all of us will be able to move mountains by the time we finish recoding. However, we must learn how to harness our energy to enable us to do that. Naturally, living life without fear was just about as empowered as one could get on this planet, but I guess I expected more. I had become highly psychic and was rarely in a situation where I could not obtain an answer to a life event. I was also able to manifest quickly and easily. But, something was missing. It seemed I had been given a pair of wings with no flying lessons.

I felt like I needed to choose an area of interest and funnel all of my newly enhanced energy into it. I wished to conquer the field of manifestation, bringing to others the skills that I was developing in that area. I wished to use soul-clearing techniques and repatterning of negative energies to removed the blocks others experienced that prevented them from moving into their manifestation power. So many on this planet do not believe they can have whatever they want. I wished to be able to teach others how to quickly and easily manifest so they could bring joyful experiences to their lives. I understood that graduation from the nine segments was only the beginning. I felt myself slip into segment eight as I viewed that obstacle and decided to have fun with it rather than feel defeated.

More about Managing Caretakers

As much as I appreciated Kryona assisting me while my astral body was away over extended time periods, I was relieved to no longer need my caretaker as the remaining work on recoding could be done during my sleeping hours. However, my friends and associates who were undergoing recoding were having a difficult time adjusting to caretakers. Julia, my roommate, was particularly unhappy since she felt flat and emotionally removed from her life, a telltale sign that a caretaker was in charge. She was enjoying her separation from her husband, and had begun to date. But, her caretaker had thrown a wet blanket over her enjoyment because she was not emotionally vested in Julia's life. My friend, Pat, had named her caretaker Rachel, and was trying to get along with her. However, she also found her emotionally vapid and particularly disinterested in working which was putting Pat behind schedule.

Both women were experiencing binges with junk food, a habit they had relinquished long ago in exchange for healthier lifestyles. However, both also confirmed they planned to endure their caretakers since the objective of twelve aligned and connected DNA strands was worth this interim aggravation.

I advised Julia and Pat to allow their caretakers to read their records because our Lightworker bodysitters were not allowed to assess our personal records without permission. Once a caretaker was allowed to access one's records, they were better able to act like the recoder in daily situations. They were also more likely to align with the recoder's emotional reactions since they had complete knowledge. Yet, they were unable to manifest full emotional reactions because they are simply replacements with no vested interest in particular outcomes.

I informed Julia and Pat that I initially resented giving up a part of myself and allowing another soul into my body. Both Pat and Julia agreed that having a caretaker was unpleasant, and they would be glad when it was over. Similar to my experience, friends noticed Pat and Julia were not themselves, remarking on their different personality or mood and the odd look in their eyes. I empathized and told them the duration of the caretaker's occupancy was thankfully brief. I came to understand that my caretaker was doing me a favor by allowing me to "disappear" for several days at a time to regain my DNA. In exchange, I tried to enhance my caretaker's third dimensional adventure by encouraging her to try a variety of different experiences. I explained how I had negotiated to be one-fourth present during Kryona's occupancy and was able to show her around town. I also had given Kryona permission to override me on anything she could do better, as long as it was in my highest good.

Attaining Twelve Strands Energetically

I have been experiencing continuous joy and elation which was a novel experience for me. It occurred to me that all of my twelve strands were aligned and fused. I had recently entered segment nine although the genetic engineers were still working on the circuitry in my brain that must work in tandem with my new strands, I felt great pride at completing this first phase of my soul contract.

Segment nine is called "Freedom from Guilt." This was when all of the twelve strands of DNA are aligned and fused. It was pure pleasure to have all of my guilt implants removed in addition to the fear implants removed during segment eight. This was the first time in my life I remembered making decisions that were not based on fear and guilt. Although I still resided in the density of the third dimensional experience, my energy had significantly lightened since I was vibrating at a higher rate. I felt like an orchestrater of experiences since I was able to move energy more quickly and easily. Strangers also noticed me as I projected an inner and outer glow from my increased vibratory rate. I was enjoying every minute of segment nine as my fear and guilt implants were removed. Making decisions and dealing with life's difficulties in a third dimensional reality were becoming increasingly easy. Even when life did not proceed smoothly, I retained a sense of complacency that everything was as it should be.

I had gone through recoding at record speed. It appeared that I would be complete with the process in five months, slightly over half the initially projected time of nine months. However, Joysia had told me that everyone would not proceed as quickly. For one thing, I was extremely motivated by my soul contract to bring the twelve-stranded energy to Earth, and I proceeded through recoding with little fear. Secondly, Joysia had allowed me to move quickly since this was a learning experience for him and his team. They wanted me to complete the process for the education of the genetic engineers. Thirdly, I was privy to Holographic Repatterning which cleared me quickly from negative frequencies that were hampering my recoding process. If I had addressed my blockages from an emotional perspective, it would have taken longer to clear them through introspection and change.

Joysia had asked me to check with him prior to using Holographic Repatterning in clearing new recoders so they would not rush their process. He said to let others know they would not complete recoding as quickly myself. I needed to honor their readiness and commitment, as well as the necessity to remove blocks. A slower rate would allow others to proceed through the segments more smoothly. I have noticed that those who began recoding after I did remained in each segment longer. They spanned multiple segments simultaneously, perhaps focusing in segment three while doing work in four and five. I have also noticed that caretakers were used earlier in the process, appearing as early as segment three, and that the caretakers were in the recoder's body for shorter time periods to minimize the discomfort on the recoder. Initially, Anu and Joysia were overjoyed to see me proceed so quickly in order to bring the full DNA energy to the planet. However, based on the stress it caused, a slower pace was planned for new recoders.

Interaction with Twelve Levels

I flew to Colorado to train with Venessa on soul-clearing. The work was fascinating and provided additional insight into my recoding experience as well as confirmation of certain elements. Knowing that my soul destiny contract involved teaching recoding, I wondered why I was so interested in learning soul-clearing. Venessa provided that answer for me the first day. She said that souls who had journeyed to Earth for service to the planet could not return to their original realm or proceed to higher dimensions if they carried dark energy from prior lives. They were not allowed to return home because it would be like breaking quarantine, essentially introducing a negative energy that would harm other Light beings living in a pure environment. Soul-clearing was the ultimate detox, clearing negative energy and past life ties with dark energy in one session.

Based on the soul-clearing Venessa had done for me several years earlier, I knew it was easy to be carrying dark energy without knowing it. For example, I had been the recipient of monitoring implants placed during ET abductions I had endured as a child. I would not have been allowed to return with these implants in my body since they maintained a link to an intrusive society who might misuse them at a later date. However, I had no knowledge of them until they were uncovered during my soul-clearing. My life had changed dramatically after that since the implants had been placed in my third eye and blocked my ability to meditate or receive psychic transmissions. Right after my soul-clearing, I began my automatic writing. Based on my practice sessions on recoders, I was aware that many with whom I worked had some type of prior dark influence that required clearing, whether it be a conflicting soul contract or a curse or a compassionate connection with a dark being or any myriad of situations.

In the evenings after the workshop, Venessa channeled an entity name Maria for our education and enjoyment. During our first evening together, after sharing some information on Goddess energy, Maria asked if anyone had questions. Of course, I always had questions these days! However, I politely sat there as the others introduced themselves by their first names and asked their questions. When my turn came, I introduced myself as Anne. Maria immediately responded that she did not recognize that name with my energy. I was perplexed, but decided to introduce myself as Lahaina. Maria said, "Ah yes, Lahaina," as if we had known each other for thousands of years. This was wonderful confirmation for me since I had never told Venessa my soul name. I proceeded to ask Maria about my twelve strands of DNA, looking for further confirmation regarding my progress. Maria said, "yes, you have twelve strands," in a tone that implied, "Don't you know this, you sweet moron?"

I asked Maria why I did not feel differently with twelve rather than two strands, and what I could tell others to anticipate through recoding. Maria explained that two strands of DNA gave a third dimensional being access to two levels of awareness and information, whereas twelve strands provided access to ten additional levels. Later, Venessa and I discussed how the twelve strands became integrated into the physical body. We saw this as a process that occurred over time since a quick integration would probably blow a third dimensional being's circuitry. In fact, Venessa surmised that those who seeded the twelve strands energetically would not carry twelve physical strands during their current lifetime. As two-stranded beings, we can carry the energy energetically but we cannot carry it in our physical bodies. We are not designed for twelve strands *physically*. However, in time, future generations will be able to physically carry the strands.

Venessa proposed that our purpose was simply to ground the energy on the planet to allow future generations of twelve-stranded individuals to be seeded. However, we also felt some integration would occur for those who carried the twelve strands in their energy field. For example, as the Earth's electromagnetic field continued to lighten in our transition from third to fifth dimensional consciousness, a soul housed in a two-stranded physical body would begin to access more and more capabilities from the twelve strands that existed in their energy field. The lightning of the energy allowed a better transmission between the physical and energetic fields. Later, I would discover that we could actually plug into the twelve-stranded energy through our endocrine system.

Based on this refined understanding of recoding and with the help of my genetic engineers, Laramus and Joysia, I rewrote the information package that I distribute to those interested in recoding. I felt it was important for those interested in recoding to understand they will not turn water into wine or walk on water after they finish the nine segments. The following channeled information was included:

"When a third dimensional being regains twelve strands of DNA, those twelve strands are attained energetically through the astral body since the third dimensional body is currently too dense to accept the higher energy afforded by more strands. This is the first step in bringing full consciousness to Earth beings. The next phase consists of plugging the twelve energetic strands into the endocrine system. Eventually, the twelve strands will be integrated into the physical body as the energy field lightens on Earth. Bear in mind, those who have energetically attained their DNA and have children will be able to seed a soul with a higher energy capable of housing twelve coiled strands in his or her physical body. In the meantime, the third dimensional life experience is improved through the gaining of access to twelve levels of awareness and information versus the current two levels. In addition to having access to greater awareness, all fear and guilt implants are removed as a part of the recoding process, again a significant improvement.

Recoding is as close as a third dimensional being can currently come to full consciousness given Earth's density. Recoding is not to be confused with ascension which allows a soul to consciously move back and forth between dimensions by activating their Merkabah. However, recoding ultimately guarantees that you will ascend since only fully conscious beings can properly activate their Merkabah for the event."

Recouping Archangel Kamiel's Support

There are seven Archangel realms who are actively involved in bringing teachings to the Earthly realm. As a starseeded soul, I had come to Earth with a specialty in a particular Archangel realm, my gift for assisting and being of service. My primary Archangel realm was Gabriel which is the realm of teaching and communication. I have always felt comfortable in teaching roles. During our training, Venessa taught us how to evaluate our receptivity to all of the seven archangel realms since we should afford ourselves assess to the teachings of the realm even if it is not from our primary Archangel. Essentially, we determined what percentage we were "open" to each particular realm. Most of my percentages were high. With Gabriel, it was one hundred percent. However, I was only fifty-three percent open in the Kamiel realm which is the realm of power.

During our lunch break, I took a long walk in the mountains outside Venessa's home. It was wonderful because I was completely alone and could hear my guides clearly due to the lack of interference from electrical power. As I walked, I asked my guides why I had been halfway shut to the power realm when it was so important for me to feel empowered in order to accomplish a significant task like recoding. They told me I had shut myself off from my own power because I did not trust it. When I had been given a great deal of responsibility and believed in my own power in the crystal city of Sedona, I had been forced to make decisions that wiped out my society. I was afraid to take my power back because it would mean I would have to, once again, be responsible for difficult decisions. My guides had been concerned I would reject the information if it had come to me first based on my rejection of my power. I cried as I realized how I had not allowed myself to flourish because I was afraid of the outcome. My guides kept asking me if I wanted my full-fledged power back, but I resisted. What if I had to make decisions that hurt others again? What if I just wasn't up to it?

Then I realized that if I did not accept my power, I would never successfully complete my work during this lifetime. I wept as I pushed my hesitancy aside and agreed to fully accept my power and the responsibility that followed. However, this time I would not be alone. I would draw on the strength of my guides and the Divine Creator to ensure I walked the straightest path to my objective, making decisions that would heal rather than harm others. I felt a tremendous fear lift from my shoulders, replaced by a feeling of calm responsibility. I knew I had cleared a major obstacle by releasing a fear I had held for the hundreds of lifetimes subsequent to that fateful experience at the Halls of Amenti in the crystal city.

Making Way for Asalaine

Venessa channeled Maria for us again. This time, when questions were asked, I focused on the whereabouts of Asalaine, my life partner. I asked Maria when I would meet Asalaine, and she told me I had to clear that "other man" from my energy field before Asalaine would appear. I was surprised because I had done a great deal of clearing to energetically release my ex-partner, Jerry. I thought I had cleared most of it. Maria said I had cleared my own issues, but Jerry had placed a "wedge" under the door leading to Asalaine. Although I thought we were done with our connection, Jerry's emotions that had originally established our union still existed. Thinking of all the energetic cleansing I had done, I wondered how those experiencing death or divorce ever released themselves from each other. There must be a tremendous number of ties that remain linked for years despite the dissolution of the relationship. Eventually, I designed a ritual to break these ties which my divorced clients deeply appreciated.

I asked Maria what I should do to dissolve Jerry's tie since our relationship was finished for this lifetime. She said I needed to use love to release it. She told me to envision a door with a wedge under it, the wedge symbolizing the energetic block that Jerry had created. Maria instructed me to connect with the Divine Creator and bring white light into my third eye, place my hand on my third eye while formulating the intent of dissolving the tie, and move my hand to my heart chakra once I stated the intent. I was then to transfer the heart energy (holding that intent in pure love) to the wedge in the door and dissolve it. Instinctively, I moved my hand out from my heart to pluck the wedge from underneath the door but Maria stopped me. She said, "You cannot use force to remove the wedge. Force creates resistance which defeats your objective. It must be dissolved with pure love." I relaxed my hand and continued to send pure love to the wedge. I saw it dissolve in a blast of energy. When I opened my eyes, I asked Maria when I would encounter Asalaine. She said, "Today, tomorrow, whenever you wish!"

During lunch, I took a walk in the mountains and asked my guides where I should go to meet Asalaine. They told me to "go to a park where people walk dogs." I laughed, saying "I don't have a dog." Their response was "Go anyway!" I was traveling almost every day through the end of August. I lightheartedly resolved to explore "dog watching" after Labor Day.

Lahaina's Translation Error

Venessa channeled Maria for us one last time on Sunday morning. Since I had all of the information I needed about Asalaine and my twelve strands of DNA, I decided to ask about my role in the Halls of Amenti. Although I was capable of receiving my own information, I liked to receive confirmation from others, especially those who had not been privy to the information I had been receiving over the last few months. Venessa was not well-acquainted with what I was doing with recoding, and Maria was a new being she had recently begun channeling. With the understanding I would be receiving a fresh and unbiased perspective, I asked Maria about my role in the Halls of Amenti.

Maria said she had to obtain permission to open the doors to the Halls of Amenti. I waited patiently as she checked, then said she could open them for just a few minutes and to ask my questions quickly. She confirmed that I was, indeed, present in the Halls at the same time I lived in third dimensional form on this planet. She also confirmed I acted as the keystone in the Halls for this project. Here is what she shared:

"I have to translate through a different energy field to make contact. You are both here and in the Halls of Amenti at the same time. This is why you have an energy translation error for your body. Your "clothes" (energy bodies) have never fit quite right, but they will within the next few months. You need to bring the two places forward through the work you do with the body, not for this place to become that place or for that place to become this place, but to combine the coded information for both places to bring two parts of the code together. It is not a soul contract for spiritual work. It is a soul contract for physical work. The body issue you work with is the translation error for the information coming through. Make your clothes fit, and this will resolve itself. You are the keystone of this particular project. But, it is a part of many great projects for Earth. It is a specific teaching you have in a different energy interpretation. It is a teaching you have brought to other worlds. It is similar to your work in the crystal city of Sedona, but it is more. It is your training for many, many, many interpretations and many worlds so you can bring a certain teaching to a different world with the same need. You have to follow your soul plan. I have difficulty reading this because of the challenge of the type of language you wish to bring here to interpret. It is the challenge of the system of the planet not wishing to expand. But, of course it will expand. The challenge is because of the lack of current expansiveness on the planet so you cannot keep the door open for great periods to bring the expansiveness in. It is time to try to ask it to open, even for a little bit, because it brings expansiveness for all. Ask, how can you open the door more, even for a little bit? Even if you open it a little, it allows the energy in. You have to set the energy vortex to allow the door to open because it is bringing information with a vibrational energy that is not accepted well by planetary Earth. So, it is a challenge. But, you have done this in other systems. You are practiced."

I was mystified by this information, hoping it would become clearer after I worked with the doors to the Halls of Amenti. After the channeling, another woman from the seminar, Kay, came over to me and said, "Do you understand what Maria meant by the translation error?" I said I did not and would she please explain it. Kay said her interpretation was that the translation error was due to the lack of heart energy in the Halls of Amenti. The energy was blocked from Earth because this planet's frequency is based on heart energy, a foreign substance to fifth dimensional realities. In fact, my greatest struggle to adapt to this planet has centered around opening my heart energy. I had spent the last ten years painfully peeling back layer after layer to get to the essence of why I came here to experience this planet.

My heart is finally open which is why I knew they would send me Asalaine, a male energy who also hosted an open heart and could support me in this venture without the need to work off karma in the relationship. Kay said if I wanted to access the Halls of Amenti, I needed to conduct a transference of energy, taking in the energy from the Hall and projecting my heart energy. Because I resided as a keystone in the Halls, I would be capable of making the transference. However, the doors would only open briefly so I had to act quickly. Over time, the doors would remain open for longer periods of time as the energy from the two locations merged.

The Path to Asalaine

My friend Wendy came to visit and do some healing work. During the course of the conversation, Wendy mentioned a man she had met walking his dog in the park near her home. They had fallen into the habit of meeting every day since she walked her dog at the same time. She spoke about his intensity, his well-developed feminine side which made him noncompetitive and an agreeable confidante, along with his familiarity with metaphysics. In fact, he was a teacher of Polynesian or Huna spirituality. He was also a chiropractor and nutritional counselor. He had recently been divorced and asked Wendy if she was interested in dating. She said, "no," she was not interested in dating but would love to be friends since his intensity fatigued her after a period of time. In fact, she was thinking of fixing him up with her friend, Sharon.

As I listened, alarms were going off in my head. I thought over and over again, "This sounds like Asalaine!" I asked what type of dog he had, thinking it might be a Terrier, but Wendy said it was a large dog and she did not know the breed. Half expecting the answer to be "no," I asked my guides if the man in the park was Asalaine, and the answer was "yes!" I sat there stunned, partially in disbelief and partially full of faith that everything was proceeding as had been promised. I said to Wendy, "You can't fix this guy up with Sharon. This is my guy. This is Asalaine!" Wendy gave me a blank stare and asked "Who in blazes is Asalaine? This man's name is David."

I gave Wendy the background on Asalaine since I had never shared my search of my perfect partner with her. When I finished the story, she expressed her excitement at being the link that would bring two perfectly suited life mates together. We debated what to do since David was interested in Wendy and was unaware of me. I knew I couldn't just call him and say, "Hey David, our guides have informed me that we are perfect life mates and meant to spend the rest of our lives together!" Wendy knew David was away for the day so we decided she should leave him a vague message on his answering machine. She called and told him she had a friend who he should call due to this friend's similar interest in the channeled Abraham material. I laughed because I had no familiarity with the Abraham material at the time and was unsure how I would back out of the white lie when David called.

I had hoped to hear from David either Saturday night or Sunday morning because I was free to meet him Sunday afternoon. I was curious to face this man who had my same grid structure. I was also very nervous, wondering if he would like me or I if would like him based on our similar energy patterns. I assumed if he was really Asalaine, we would have an intense mutual attraction. I left the house at 11:00 a.m. Sunday and missed David's call. He called at 11:05 a.m. His message said he was responding to a call from his friend, Wendy, and would be happy to talk to me on Monday when he returned from a conference he was attending.

In an effort to prepare him for our conversation, I called and left him a message. I told him I was unfamiliar with the Abraham tapes, but I had received information from my guides that we should meet because we shared common business interests. It's awkward knowing more than the other party. I was entering the situation with a certain set of expectations. I did not wish to create the same set of expectations for David, perhaps making him nervous or causing pressure. I figured the pretext of working together would suffice for meeting, and we would know in the first fifteen minutes if it was more than that. I awaited his next call.

Asalaine and Lahaina

David called my office at 10:30 a.m. My assistant answered the phone, then called to me, "Dr. David is on line one." My heart leaped into my throat, and I started sweating. I thought how silly it was to have such a schoolgirl reaction, but I could not seem to control my nervousness. What was I going to say to a life mate I had never met? At least, I knew per Wendy that he was tall, physically fit, and nice looking. He was rather disadvantaged because he knew nothing about me.

I picked up the telephone and said, "hello," and I repeated what I had left on his message recorder, that my guides said we had work to do together and should meet. I explained how I had been told months prior that I would be meeting a chiropractor in his fifties who lived in Kansas City, was involved in healing, was well-acquainted with metaphysics, and, of course, had a dog. The soul name of this man was Asalaine. I asked if he had ever heard that name in relationship to him. He said, "no," and I could hear the lack of warmth and trust in his voice. I felt like I was using some cheap, metaphysical pickup line in the genre of "what's your astrological sign?" His tone suggested that he felt the same way. I struggled with what to say next to break the ice and I heard my guides say, "Tell him your soul name. Say that you're Lahaina." I took another shot and asked him if he had ever heard the name "Lahaina." There was an immediate adjustment of the energy as I felt him connect. He said, "Who are you? I just got a hookup. You know that rush you get when you gain immediate insight? That's how I feel."

David said he had heard the name, Lahaina, and it meant something to him, something from the distant past. I asked if he was thinking of the town in Hawaii? He said, "no," but that he taught Polynesian metaphysics, thinking that might be the connection. Later, we would determine that one of our happiest lifetimes together was when we lived as two Polynesians in the Hawaiian Islands. During that lifetime, I had used the name "Lahaina." I asked David if we could meet for an hour and explore if there was any credence to our "business" connection. He agreed, and we struggled to match calendars. I was leaving town for a week, and he was busy with his patients and his weekly radio show on alternative medicine. We settled on meeting at 3:00 p.m. the following Saturday. We were to meet after his show at a restaurant in a nightclub section of Kansas City. He kept stressing he only had thirty minutes to an hour free due to his busy schedule. Clearly, he was hedging his bets, and I did not blame him. I was disappointed to wait another week, but knew either he or I had unfinished business to resolve energetically or we would have connected sooner. I hung up the phone and prepared to wait.

Severing Final Connections with Jerry

I had to call Jerry due to some unresolved finances regarding the house. The call led to a series of emotional conversations throughout the day that resulted in the final resolution of our breakup. It was emotionally difficult but also very cleansing. We both had felt we were meant to be together forever, that ours was a divine soul complement relationship that would transcend time. It had been difficult to resolve that soul mates were not necessarily supposed to be together forever. In fact, as my guides explained later, soul mates were often characterized by relationships full of passion because the two individuals had much karma to resolve from prior lifetimes. Had the relationship not been passionate, most couples would not have remained together through the clearing of karma. Jerry and I had been soul mates and had completed our karmic connection. Now, we were free to find our life partner.

Somehow, through the course of our conversations that particular day, we were able to put our lack of understanding to rest. We could honor the relationship we had together and know the love would not die, while also honoring the fact that we were not meant to be together as life mates. There was no negative emotion left, only a feeling of peace. I knew by the end of the final emotionally intense conversation that I had completely let go of him and vice versa. I also knew that this was why I could not have met Asalaine the previous weekend. Sometimes we need to clear and come to peace with certain aspects before we can move forward, especially when we are so emotionally bound.

Later in the day, Wendy called to tell me of her conversation with David in the park that morning. Apparently, he had not entirely bought my "business" excuse for getting together. I guess when an unknown female calls and claims her guides suggested a meeting, that carries a hint of a come-on. Later, David would tell me that some women had tried this tactic in the past without success which is why his suspicion was raised. According to Wendy's most recent meeting with David that morning in the park, he was more interested in asking Wendy about my personal rather than professional habits.

Wendy told David that he "owed her" for getting us together because I was the "sexiest" friend she had. I was dismayed by this promise because I did not feel I fulfilled her description. I hoped I lived up to her portrayal in his eyes when we finally met. She also said David had shared with her a weekend experience he had with his ex-wife. Apparently, she was also at the conference he was attending. When he saw her, which was prior to receiving my message, he looked at her and felt totally disconnected from her energy. He felt released by the emotional ties of the relationship. As he was driving home from the conference, he felt a lifting of the heaviness he had carried for months. He felt liberated and began singing due to his incredible feeling of joy and freedom. He told Wendy he knew he was ready for someone else, the ultimate partner he had never had. When he returned home from the conference, the message light was blinking on the machine and my voice emerged. He had been free from emotional connections for exactly forty minutes! Later, he marveled at the synchronicity of endings and beginnings.

I had mixed feelings about David's realization that my call might mean more than business. On the one hand, I felt it was better that both of us enter our initial meeting with equal intent and with our eyes open. On the other hand, I felt the playing field was being loaded with obstacles that created pressure for both of us. Nonetheless, I felt the week drag until we could finally meet and satisfy our mutual curiosity.

Graduating from Phase I of Recoding

Today was the day of my follow-up pap smear. I had made and canceled the appointment so many times, due to scheduling conflicts, that I knew my guides did not want me to go. Yet, I felt it was necessary to place closure on the situation. I entered my HMO and sat down to wait. I waited for nearly a half hour, realizing I was going to miss my second appointment if they did not see me. I approached the front desk, asked how much longer, and was told they were ready to see me. I returned to my seat and waited an additional ten minutes. Finally, I decided this was not meant to be and returned to the receptionist, asking her to return my plastic membership card.

The receptionist said, "One moment please," and went to get the card which was with my medical records. When she returned, a nurse was with her. The nurse asked me to step into the back to speak with her for a moment. I entered an examining room and asked her what she wanted. The nurse said she was sorry for the delay but there was an error on my chart. The appointment had been scheduled for a pap smear when I needed a kolposcopy. I said, no, I had changed the appointment to a pap smear because I was not receptive to someone scraping my uterus. The nurse said the pap smear would not be helpful because it would simply reveal what it had the first time. Knowing I was cleared from whatever caused the initial problem, I told the nurse I would agree to have a pap smear but not a kolposcopy. We went in circles for several minutes, she insisting I needed a kolposcopy and I wanting a pap smear. I finally asked her to define "questionable pap smear" since no one had bothered to tell me. She said she was not "at liberty" to reveal that information.

At that point, I was fed up with the situation. I had given authority to medical professionals for too many years and no one was going to tell me they could not inform me of my health until they could conduct their next intrusive procedure. The patient is rarely allowed choices. I thanked the nurse for her time but said I had another appointment and needed to leave. I would see them in March for my annual pap smear.

When I reached the lobby, I thought to myself, am I crazy? Am I risking my health and my life over this? Then, I decided it was the only path I could follow. I was not interested in surgeries, medications and nurses with half answers. I would rather die gracefully if my time had come. I called N, my next appointment, and told her the reason I would be late. She began laughing, saying, "Anne, you just graduated from recoding. You've cleared the nine segments. I see a whole room of beings cheering around you because you took back your power. Congratulations!" Sure enough, she was right. I had been too busy fuming to notice that it was uncomfortably hot around me as I felt their presence. I had done it! I was the second person who had completed this version of DNA recoding! I told N I would be right over.

I had several questions about the nine segments that I wanted to clarify with Joysia before publication. I also had several personal questions, including feedback on my upcoming meeting with Asalaine. N began channeling Joysia who answered my questions on the writings. The final version of the nine segments, which has been distributed to many instructing them how to begin recoding, is located in this book. At the end of the session, I asked Joysia to help me understand my role in recoding beyond teaching it. Joysia said he would read a segment of my soul contract from the Akashic Records. He said I would use portions of Holographic Repatterning and soul-clearing to assist others in clearing their emotional and psychic pasts. This would enable my future clients to remove blockages to their soul growth which would either prepare them for recoding or expedite their recoding progress. I would use my psychic ability to provide information for repatterning and soul-clearing, but I would not specialize in psychic readings. He also said it was never meant for me to be a full-time repatterner or soul-clearer, that I had learned those methods in order to support my work on recoding.

Later, I would realize how accurate that information was as I tapped my repatterning and soul-clearing tools to develop the Plug-in process of the twelve DNA strands into the endocrine system and the Balancing Polarity ritual. At the time, I was pleased and not surprised by this information since I had witnessed incredible healing with repatterning and soul-clearing. The negative drain that people carry around for lifetimes can be cleared instantaneously, resulting in major improvements in their current existence. I was happy to follow this path, realizing it was the path I was already on. When was I going to get it through my head that I was always exactly where I needed to be?

Finally, I asked about Asalaine. I thanked them for bringing him to me. They said, "Dear one, you drew him to you. We simply helped you clear the blocks." I asked if I needed to know anything prior to meeting him. They said, "Just be accepting." This is good information for a person whose dominant Archangel realm is Gabriel because we tend to not only be self-critical but critical of others. We are known for our perfectionism and must constantly battle it. I promised myself I would be as accepting as possible, wondering what was so unappealing that I had to be accepting. Maybe, it was his preppy style of dress since I was averse to formality, adopting a casual style of dress after several years of self-employment.

Meeting Asalaine

Finally, it was Saturday. I had to fight from calling David to cancel the whole meeting because I was so nervous. I have never felt confident in my appearance, and on top of my own feelings of insecurity, Wendy had set the stage by telling this man I was one of her sexiest friends.

I went to the designated restaurant, and sat at the bar. I purposely arrived early because my legs were like jelly, and I did not trust David watching me as I entered. I ordered a cup of coffee and sat while my stomach churned. He had told me he wore his hair in a ponytail so I spotted him immediately as he entered the restaurant. I was instantly drawn to his eyes. They were incredibly intense and full of passion. He was not what I pictured, yet he was more than I imagined. He was a lit light bulb. He was so filled with life, something I had missed intently when living with Jerry. Indeed, despite the long hair, he sported the conservative style of a white golf shirt and Dockers. However, the contrast of the hair and the preppy clothes was very appealing.

We sat and talked, both having *plenty* to say. About an hour into the conversation, I wondered what he thought of me, if he was attracted to me, or what was going on in his mind? I was a nervous wreck, hoping if I kept talking that he would not have to leave. Abruptly, he asked me if he could touch my hair. I hesitated, then said, "Sure," wondering what he was thinking to himself. He reached over and touched my hair, and I felt a surge of energy pulse through my head. He said something about being fascinated with curly hair and wanting to touch it. Whatever! At that point, all I knew was he was elegant, charming, articulate, intelligent, passionate, generous, and full of integrity based on our conversation and my observations. I did not know how he felt about me, but I knew the way I felt. I felt with all my heart that this was my life mate!

David began describing his house to me, telling me he wanted to show me his classroom. He taught metaphysics in a classroom on the third floor of his one hundred-year-old home located in a historic part of Kansas City. His home sat on an energy vortex he called "the well" that shot the energy up through the core of the house, magnified by the roof which was a perfect pyramid. It was a tremendously powerful healing energy. As he spoke, I realized a friend of mine had taken his classes and that she had once described his house to me. I told David I thought we would have met eventually since we had a mutual acquaintance. I said I would love to see his classroom but needed to pick up a book at a bookstore close to the restaurant. We decided to swing by there on the way to his home. Guess who was in the bookstore when we arrived? Our mutual acquaintance—talk about synchronicity!

She looked so surprised when she saw us, exclaiming "I didn't know you two knew each other!" We just laughed at her look of surprise. Our mutual amusement already made me feel like David and I were a couple. After chatting for a while, we left for David's home.

David showed me around his house. By now, my head hurt from so much energy moving between us. My stomach was tight because I was still nervous. I still had no idea what he thought of me since nothing had been said. The house and the classroom were beautiful, and I could see how powerful David was from what he had created. He asked if I wanted to sit in the well and experience the energy. I said, "yes," and sat on the bench directly above the energy source. The energy was phenomenal, so clean and pure. It tickled me as it went through me, reminding me of that feeling right before an orgasm. I blurted out, "Have you ever made love to anyone in the well? It would be fantastic." I might as well have said "When are we going to make love in the well?" Surprisingly, he said "no," that he had never met the right person, suggesting that finally he had. After my embarrassing comment and the tour of his house, I said I probably needed to leave since I had overstayed my visit. Remember, he was this "busy person" who only had the maximum of an hour allotted in his schedule to meet me on Saturday. We had already been together for three hours. He said, "You're not leaving because I need to take you to dinner." No problem, although I didn't know how my stomach was going to handle eating. I assumed from his comment that he was enjoying my company as much as I enjoyed his.

We went to a local restaurant. There had been no silence between us, and we had already spent three wonderful hours together. I knew I had loved him for many lifetimes and felt very comfortable with him despite my nervousness. But, I had no idea what he was thinking or feeling about me, and was afraid to ask. As we sat looking at the menus, he looked me straight in the eye and asked if I was aware of the heightened energy between us. I told him, as I looked into his eyes, that I was very much aware of it. He said, when he touched my hair in the restaurant, he felt an electric shock. He had felt impulsed to touch my hair because it surrounded my head like a halo and made me look like an angel. He also remarked he was exceptionally nervous and excited. Well, at least that made two of us.

David asked if I knew why there was so much energy between us. I told him that I hesitated to share information that might cause unnecessary pressure. He said he was already overwhelmed, so I might as well tell him. I told him the information I had received from my guides. I tried to downplay the fact that I had been working on manifesting him since early May. I explained how we had the same grid structure, that we aligned with each other better than anyone else and that no one on the planet had the same energy patterns. Talk about shocking someone on the first date! David took the information surprisingly well, commenting that my explanation clarified the way he was feeling about me which was equally intense.

After dinner, we returned to his home and walked his dog, then sat on the swing on the front porch and talked. We still had not run out of things to say. It was such an odd feeling. I felt like I knew this man my entire life, but I really knew virtually nothing about him. Deep in my heart, I also realized this was everything I had ever wanted and we would be together a long time. I wondered why Joysia had told me to be accepting since I felt very accepting of this man. There was no trepidation, only relief that I had finally found him. I wondered if he felt the same way about me?

David began talking about how he had been looking for the perfect partner his entire life. He knew exactly what he wanted in a relationship and, even during his multiple marriages, realized he had not yet attained it. He told me how he had created a list of the perfect person for him that he reviewed every day. He had also been in the process of manifesting someone. I laughed when I realized we had both been creating each other at the same time. David also shared his feelings regarding closure with his most recent ex-wife and how my phone call awaited him when he got home that day.

I asked to see his perfect partner list. I thought it was brave of him to agree to show it to me. It is amazing how much we are willing to reveal about ourselves to total strangers, especially after both of us had been so recently hurt after opening our hearts up to relationships that had not succeeded. But after all, we were not strangers. This was Asalaine who had come into physical form to be with me. And, he had been looking for me as much as I had been searching for him! David returned with his list, and I reviewed it. The list described me perfectly with one possible exception. He wanted a "sailor," and I had only sailed once on a tiny boat. I said I did not know if I were a sailor, but he said he already knew I was.

It was getting dark so we went inside. In the living room, he reached out to kiss me and it was so intense, so immediate, and so overwhelming that we both had to sit down. Then he took my hand and led me upstairs. That was the beginning of forever, and we both knew it.

Summary: August Recoding Lessons

- Segment six, "Owning Your Power," and segment eight, "Releasing Fear," both involve manifestation. However, segment six focuses on teaching us to use our own power to manifest rather than relying on requests to spirit guides. Segment six is a recognition of the recoder's power as fully capable of manifestation. But, the recoder is somewhat hindered in segment six to manifest based on the number of unfused DNA strands. Then segment seven is experienced which enables us to test new manifestation skills in light of illusionary obstacles. At this time, we are tested in terms of letting go of fears that would materialize less than positive outcomes. Segment seven is tough because one moves from the realization of self as the power source to running the gauntlet to test this power.
- Once a recoder has mastered the ability to unveil illusion and allow events to work in their behalf, they are allowed to enter segment eight which clears the illusions from segment seven and allows manifestation. Segment eight is

also when one receives at least ten of their twelve DNA strands so manifestation becomes quick and easy.

- Manifestation must align with one's soul contract since all of the power from the twelve DNA strands will be focused on fulfilling those contracts. This is when one uses their new power to become adept at a particular line of work. Recoders will not be using their new power to create youthful appearances or turn water into wine. However, when one is in flow with their soul contract, they are in a position to manifest the rewards of the third dimensional experience.

- In segment eight, all of the fear implants are removed and in segment nine the guilt implants are removed. One is finally able to experience life in joy and love rather than fear and guilt with all of the benefits afforded by that viewpoint. Fear and guilt implants were used to ensure one learned their Earthly lessons well since these were key components of the third dimensional experience. However, they no longer serve a purpose after recoding.

- Caretakers will do a better job of "house-sitting a recoder's body if they are given permission to read the recoder's tapes. This ensures that a caretaker will react to unplanned situations similar to the recoder they are trying to replicate. They are also more likely to align with the recoder's emotional reactions since they have full knowledge of the recoders thoughts and feelings. They are unable to manifest full emotional reactions since they are still replacements with limited third dimensional experience and no vested interest in any particular outcome.

- Segment nine, "Freedom from Guilt," is the crème-de-la-crème of the recoding experience. One now has twelve strands of DNA aligned and fused, and the guilt implants are removed. This is the first time in this lifetime we have experienced life on Earth without fear and guilt. Imagine the full capacity of one's energy without fear and guilt! Although we still reside in a dense body, our energy has lightened significantly since we are vibrating at a higher rate. This rate will continue to heighten as the Earth increases its frequency, enabling one to move energy quicker and easier.

- Anne underwent the nine segments of recoding in record speed since the genetic engineers and the Nibiruan Council wished to learn from her third dimensional experiences. However, others will not experience recoding with such speed since it causes pressure on everyone involved, both recoders and genetic engineers. Now that the twelve-stranded frequency has been grounded in the Heart of the Dove (the Midwest area between Kansas City and St. Louis) which is the entry point for the fifth dimensional energy, Anu and Joysia are slowing the pace.

Chapter Six

New Beginnings

The Formal Union of Asalaine and Lahaina

After undergoing recoding, I have a growing metaphysical practice, my ideal life partner, and I am living my life without fear and guilt. I feel like I float in bliss each and every day of my life. There is so much magic around us if we allow it into our lives. When I tell people the story about how I left fear and guilt behind or how I met David, they either have two reactions. They decide if it can happen to me, it can happen to them, and they ask me how to go about moving in the same direction. Or, they determine that I am the luckiest person in the whole world and that something like that would never happen to them. Guess which of those two types of individuals are ready for recoding?

David and I are very proud of the creation of our relationship. Everything is easy and perfect when we are together which is what Anu and Joysia promised me from the beginning. We are totally in love and, while not co-dependent, feel a piece of us is missing when we are not in the presence of the other. The love is powerful yet gentle, supportive without being consuming, and sharing instead of needy. And, as promised by Anu and Joysia, physical love between twelve-stranded beings is about as close to the experience of Divine Creator love that one can attain on this physical plane. We met on August 24, 1996, and we were happily married a year later on August 24, 1997. Even with seven previous marriages between us, there is no doubt that this was the one we were preparing for, the one we always wanted but had not yet found. We cherish our sacred union and thank our earlier partners who helped us become what we are for each other. We foresee many blissful physical years together and have pledged our union for eternity.

As we discovered soon after meeting, David was already in segment seven of recoding. When I shared the recoding information, he recognized the segments he had already experienced. We never discovered how he initiated recoding, but assumed he had completed the work prior to walking in. Apparently, Asalaine was the oversoul of David so the walk-in process was simply a transference of more of Asalaine into third dimensional form, similar to how I was becoming more and more of Lahaina. This transference began on his birthday in February, 1996 so by the time I met him in August, he had nearly completed the transition. David insists that he is still David rather than a transference of the Asalaine energy.

However, most of his prior friends do not recognize him because he has changed so dramatically. His personality is much softer and more low key than it used to be. He used to wear navy blue blazers and ascots, but now feels more comfortable in jeans and a T-shirt. When I met him, he claimed he was comfortable with his life and would never leave his home or his daily routine. By June of the following year, he had sold the traditional house he had spent years renovating à la *Colonial Home* in the historical section of Kansas City and moved into my contemporary southwestern home in the suburbs. He had also stopped practicing chiropractic in order to devote his full-time attention to nutritional counseling (sacred body) and teaching metaphysical classes with me.

I have yet to win the lottery. However, I earn whatever money I need effortlessly, and I expend a lot less energy doing it compared to my workaholic habits of the past. In fact, I earned enough money the following year to pay Jerry for his share of the house *and* pay off my mortage. Besides, now that I have learned how to manifest the desires of my heart, I expect to win the large amount of money that I have always felt was coming to me. I have cleared and relaxed my attachment, and am now in a state of joyous anticipation. I desire to use this money to expand my metaphysical practice and to experience more freedom in my life. I now realize that I have dramatically changed my life in a very short time, and have become the creator of all of my experiences. I am specifically creating my life to fulfill the desires of my heart and I am anxious to teach these valuable techniques to others. Knowing that I have learned how to link with my guidance system, and use my personal skills to vibrate with the abundance of the universe, I now know that this abundance is within me!

At this point, there are many individuals who are working on recoding, some more assiduously than others. I have found that recoding does not resonate with everyone. There are some who find it to be the solution for which they are searching, and there are others who find it to either be of little interest or to be too difficult once they start. Those who have undergone recoding will attest to major shifts in their life, not only from a personal perspective which often feels like a major upheaval, but also from an enlightenment perspective. These people are operating with greatly reduced fear and guilt in their lives and have telepathic abilities which give them access to an array of information that assists themselves and others.

I thought I had completed recoding when I finished the nine segments. However, I discovered that there were two more segments of recoding. Initially, I was disappointed because I liked feeling a sense of completion about finally retrieving my missing DNA strands. But, I also understand that our soul evolvement is never finished. Plus, I am sure that even if Anne were to reach completion on Earth, there are more expansive experiences for Lahaina to undergo in other dimensions.

Unfortunately, N and I needed to part ways after the nine segments of recoding. Her direction, via Anu, became significantly different from the information I was being given by Laramus and Joysia. Although I wish her the best of luck spreading her version of the DNA recoding information, I have felt compelled to honor the information my guides sent me regarding the final two phases. They have always provided me with guidance for my highest good.

Phases II and III of DNA Recoding

According to Joysia and Laramus, the second phase of recoding is called the Plug-in Process. It consists of balancing the endocrine system which is the window to Divine Spirit. Once the endocrine system is balanced, the twelve DNA strands can be plugged into each of the endocrine glands thereby linking the energy in the astral body with that of the physical body. I spent the last two weeks of 1996 creating the balancing modalities for the Plug-in, and I have tested them on many recoders. Everyone who has undergone recoding feels even more in touch with his or her power after the Plug-in process.

The third phase of recoding was provided to me nearly a year after I completed the first two phases. It involves balancing one's polarity for unifying in the midst of a planetary experience whose blueprint dictates the push/pull of opposing forces. These forces are called "polarity," e.g., good/bad, right/wrong, left/right. All of the phases of recoding are outlined in the following sections, providing guidelines for anyone to follow who desires to recode.

Since obtaining the information on phases two and three, I have also received much confirmation that the process works, both through my own experiences and through my clients. For example, several months after I completed phase two of recoding or Plug-in, I attended a spiritual workshop at Venessa's. One of the exercises consisted of using a pendulum to dowse the percentage of fear, guilt, and pride in our mental body since we were working on eliminating those negative frequencies. I used my pendulum and dowsed zero percent for the amount of fear and guilt residing in my mental body. I assumed these numbers to be incorrect since others in the class were obtaining percentages in the twenties to forties. I asked the teacher to check my numbers since they were apparently incorrect. Maybe, my ego had overridden the results! She verified that I held no fear or guilt in my mental body and asked what I might have done to clear myself. I told her about the recoding process which was supposed to remove fear and guilt, somewhat amazed yet not amazed that what I had undergone the prior year had worked so effectively.

During that same workshop, we also conducted an exercise to balance our endocrine systems. My teacher, Venessa, believes it will become the primary operating system in our body in preparation for the ascension from the third to the fifth dimension in a *fully conscious* state. She explained that the endocrine system was in the process of replacing the immune system as the body's key system since we could not house a "fighting" system and be balanced. In other words, the immune system used warrior energy to combat invading germs, an energy inappropriate to the balance needed for ascension. Again, my percentages were different than the others in the class. Whereas the other students had blockages that created a sixty to ninety percent efficiency factor, I was open in all of my endocrine glands by ninety-eight percent or more. Apparently, when I balanced my endocrine system during the Plug-in process, I had removed the majority of constrictions. This was further confirmation that the process worked.

In addition to my own experiences, I work with clients who continually tell me how their lives have changed due to recoding. The following quote is typical of the kind of feedback I receive: "For a spring in your step and a smile on your face, I have to say that Holographic Repatterning has given me back many things that I almost belived were gone forever. The feeling of complete confidence in what I do or say is wonderful. I now know that there are no outside influences distorting my reality and I get consistent feedback from friends that there is a smile on my face or that I am looking great or that I seem so very confident or sure in what I am doing. It's great! I have encouraged many friends to check it out and the ones that have spent time with Anne I can see a vast change in their lives for the better." — Julie Wells, Kansas City.

The recoding experience has brought people joy and peace of mind. Although they are actively participating in life, they feel comfortably removed from the negativity that affected them in the past. Abundance comes more easily to these recoders and, when there are stumbling blocks, there is little reaction or resistance which makes the barriers easy to overcome. My clients who know each other also remark on how personalities have changed. Those who led more passive lifestyles are more "on," grabbing more as they move through their experiences. Those who were directive have become more easygoing as they have learned to enjoy the flow, knowing they direct their destiny by manipulating their energy rather than feeling like they need to exert control.

For each of you, I hope you decide there is magic for the taking, and your experience on this planet is the grandest experience in the universe. Live it, enjoy it, maximize it, revel in what you create. And remember, you are only here on temporary assignment. One day, you will return to your original place of origin or you will move on to new worlds full of rich experiences. Feel proud knowing that your tenure on Earth was enriched by what you created while you were here. Perhaps that will include bringing the twelve strands of DNA to this planet. Perhaps it will be something else that is equally glorious. Whatever it is, it will be perfect for you.

Love, Anne (Lahaina)

Appendices

Appendix A

- Phase I -

Summary of DNA Recoding

Definition of Recoding

Recoding is the process by which a being of third dimensional human origin who has two strands of DNA energetically recaptures the twelve strands that were ultimately intended for them. These strands were tampered with eons ago by those who wished to use third dimensional beings for their own purposes, thereby limiting their abilities. The reestablishment of this complete energetic circuitry enables a soul to have access to twelve levels of spiritual, emotional, physical, and mental awareness and information versus the two levels afforded by two strands of DNA. Access to twelve levels results in expanded consciousness in third dimensional human form. This manifests in characteristics such as enhanced psychic abilities like clairvoyance and clairaudience, memory retention of past physical and nonphysical existences, consciousness of current simultaneous soul existences, and access to the Akashic or soul records.

Many think they are currently experiencing recoding. Please understand that the only beings undergoing recoding are those who are working with their spirit guides or genetic engineers to deactivate their DNA implants. These implants were put in place by the Pleidians to demagnetize ten of the twelve strands of DNA thereby preventing them from aligning. In order to control you, they created the two-stranded beings. Deactivation of these implants has to occur in order for the twelve strands to recoil and fuse, in energetic form.

When a third dimensional being regains twelve strands of DNA, those twelve strands are initially attained energetically through the astral body. The third dimensional body is currently too dense to accept the higher energy afforded by more strands. You would not care to reside in your body if it housed twelve strands of DNA because it would be too painful. It would give you more amplified energy than you could handle. Integrating the twelve strands into the astral body is the first step toward bringing expanded consciousness to human beings because it establishes a higher frequency on the planet. The second step involves plugging the twelve strands of DNA into the liquid system, which is the endocrine system. The third step is comprised of attaining balance and acceptance amidst the mass consciousness of polarity (e.g., good/bad, right/wrong, left/right) by renegotiating one's current polarity contracts for unification contracts.

If you are reading this information, it is highly likely you have agreed through your soul contract to be of service by bringing this higher frequency to the Earth plane. Although your physical vehicle may not house twelve DNA strands, access to twelve levels of information versus the two you currently access will exponentially improve your life since you will have more data with which to work. In addition to having access to greater awareness, all fear and guilt implants are deactivated as a part of the recoding process, again a significant improvement to your current life experience in the mass consciousness that is driven by these emotions. Originally, these implants were intended to enhance your learning experiences on Earth. However, they are no longer of value to a being with expanded consciousness. Imagine, a life of choices based on joy and love rather than fear and guilt. You are virtually guaranteed a better quality of life!

Once you have completed the nine segments and recaptured your twelve DNA strands, you will be able to funnel the energy from the additional strands toward the fulfillment of your soul contract. All the power of your twelve strands is used to fulfill your contract. It is not used to make you look young again or make you walk on water. Remember, you have not been chosen casually. If you are interested in undergoing recoding, you have a very important soul contract to fulfill for the benefit of the planet in addition to seeding the twelve strands. You will become adept at the work you were meant to perform during this lifetime as you master your particular area, be it healing, communicating, teaching, channeling, manifesting, or clearing energy. Your heart will resonate when you perform this work, affirming the correctness of your actions. This is when your financial rewards will expand and you will glow with good health because you are working in bliss, aligning your daily actions with your heart energy. All the wonderful things that the third dimensional experience has to offer comes through the fulfillment of your soul contract.

Recoding is as close as a third dimensional human being can currently come to full consciousness in physical form given Earth's density. However, recoding is not to be confused with ascension which allows a soul to consciously move back and forth between dimensions. Recoding does ultimately guarantee that you will ascend since only fully conscious beings can do so. The time frame for ascension will vary, depending on your individual soul contracts.

The How-Tos of Recoding

There are many ways to recode depending on the contract of your particular soul group. Bear in mind that the following process has been provided by non-incarnate members of the Nibiruan Council to assist anyone who resonates with this particular method. These beings are karmically involved with assisting third dimensional beings due to their past involvement in tampering with your DNA. Those of you who do not resonate with this process but are interested in recoding will most likely find another path more suited to your needs if your contract involves recapturing your twelve strands of DNA.

Prior to beginning recoding, it is necessary for you to accept your "soul contract." This is your soul's purpose for incarnating on Earth at this time as an indication that you, as a Light Being, intend to use your newly acquired powers for the good of all. It should not be difficult for you to accept your soul contract since it is your reason for being here at this time and will resonate with you. If you are not already aware of your soul contract, you will be given as much information as possible from your guides. If you are unable to communicate consciously with your guides at this time, ask a psychic counselor to help you. However, some information about your soul contract may not be shared since it will be meaningless until you achieve greater awareness which will assist your level of comprehension. This information will be provided by your guides when it is deemed appropriate.

Once you accept your soul contract, your formal request for recoding can be made to the Sirian/Pleiadian council who signifies recoding approval by changing your soul status in the Akashic records. These are the fifth and sixth dimensional beings who are most actively involved in upgrading Earth to a higher dimensional frequency. They have sent many starseeds to assist in this process of which you may be one. When reading your formal request, prepare by taking a warm bath to which you have added one cup of baking soda and one cup of sea salt to purify your system. Then, light a white candle and read the following request aloud. The format for the *formal request* is:

- "To the Sirian/Pleiadian council and all beloved brothers and sisters of the Universe, to Prime Creator who created all things in all people, I am requesting of my own free will access to full consciousness through the recoding of my DNA and I am ready to begin at this moment."

You may use a pendulum or kinesiology to dowse to find out if you received permission to recode from the Sirian/Pleiadian Council. If you do not know how to dowse or are not yet able to receive psychic transmissions, you can either purchase a book on dowsing or contact a psychic counselor. Or, you can contact Anne Brewer at InterLink (*913-722-5498*) or at the Internet website *www.interlnk.com* to schedule an appointment to determine acceptance and what blocks, if any, stand in the way of your acceptance that need to be cleared.

Recoding will require much commitment. Since you will need to rid yourself of the majority of your toxic energy, mostly in the form of anger from this life as well as past lives, you may feel worse before you feel better. In fact, there may be physical discomfort during some of the work depending on how much negativity you have cleared prior to beginning recoding. But, if you persist, you will acquire the great joy you have desired for many lifetimes on this planet. Bear in mind, you will not be approved for recoding unless you have released the majority of your negative blocks since you would endure discomfort if these blocks existed in your body when recoding begins. Most of you have been working on this for years, so you will be relatively clean when recoding is presented to you. We compliment you on your efforts since the release of negative blocks is generally unpleasant, and your society provides many options for avoiding a complete examination of yourself in the form of television, alcohol, drugs, money, work, and other addictions. It is important for you to remember that the permanent release of toxins in your body depends on the release of emotional blocks since releasing *physical* blocks does not endure. When emotional blocks remain present, the physical body might rid itself of anger one day but express it again the next day.

The following methods are recommended for releasing negative emotional patterns and toxins in preparation for recoding. These should be explored in addition to asking for the implants referenced in the Kryon material to eliminate all your past Earthly karma. This will automatically release blocks from your emotional body. If you need more information on the Kryon implants, read the information channeled by Lee Carroll.

Some of the recommended emotional cleansing methods are: energy work like Reiki, attunement, or craniosacral therapy, acupressure or acupuncture to attune the chakras, toning to balance and clear chakras, retinal therapy or New Decision Therapy, Holographic Repatterning, reflexology or massage, liver cleanses for releasing deep levels of anger, activating crystals in the crown chakra, chiropractic therapy, stretching disciplines like yoga, and transformational breathing

It is also beneficial to eat as little meat, fowl, and fish as possible to cleanse your physical vehicle. These foods interfere with the lightening of the body. If you eat meat or fowl, eat the antibiotic/hormone free/free range versions that are sold in health food stores. Remember, animal flesh creates a heavy energy in your system while recoding is attempting to lighten your energy field.

If you are reading this information and it is resonating with you, you are probably a starseed. This means you have already existed at higher dimensions, probably fifth or sixth. For you, recoding is a process of remembering what you agreed to forget when you entered a third dimensional physical body. You are simply activating the codes already residing in you. In fact, all starseeds have codes in them that impulse them to recode. This is why the information resonates so strongly with you. The only "new" information will be how to transform the human body you are inhabiting to a higher state of consciousness.

Recoding is not a completely defined process. Your guides are learning along with you since few third dimensional beings have moved to a more fully conscious state from dense physical form (as distinguished from enlightened beings like Jesus who chose to enter a third dimensional form without any implants). You will be assigned a genetic engineer during recoding to assist you. This is the new guide referenced in the Kryon materials. We suggest you communicate regularly with your genetic engineer since you will be undergoing changes that you might not understand. At times, you may feel like you are slightly crazy, but we assure you that you are not. As you move through segments three and four of the nine segment recoding process, you will be able to create your own form of communication with your genetic engineer since your psychic channels will open. However, until that time, use the skills of a psychic individual to keep communication open. Rest assured that your genetic engineer will love and care for you and will allow no harm to come to you. In fact, you will be impressed by the protection you receive during this process, and you are promised an outcome that will bring enhancement to your life.

Although there are nine segments to recoding, it is not an entirely sequential process since you might finish something from segment four then return to segment one then go to segment two then back to segment one for refinement. The nonsequential aspect of recoding is due to the fact that the codes for altering your DNA are scattered throughout your body. While asleep, your astral body travels to meet with your genetic engineer, just as you travel now during sleep to work with your spirit guides and teachers. During astral travel, changes are made in your astral body then integrated into your physical body. Sometimes, in the transition from astral to physical body, something from work on a prior segment is misaligned and needs realignment, or a code located in one part of the body is opened which causes a code in another part of the body to react.

When you begin to recode, you extract yourself from the pool of mass consciousness of fear and guilt that constructs reality for this planet. You are part of the recoder pool of consciousness, one based in love and joy. As you expand your consciousness and continue through the nine segments, the recoder consciousness grows in strength, creating a new pool of consciousness toward which Earth's mass consciousness can strive and attain. For this reason, it may be challenging for you to maintain intimate relationships with anyone who is not going through recoding since you no longer exist in the same pool of consciousness. You will particularly notice this when you finish segments eight and nine since all of your fear and guilt implants are removed at this time. How could you enjoy spending time with those filled with fear and guilt when you no longer experience those emotions? If fear is present in your physical vehicle, it can actually undo some of the work that has been done and keep you from your life mate. It is important throughout recoding to have strong faith and keep as much fear as possible out of the body until your fear implants are removed.

Some recorders will feel inclined to leave energetically unbalanced relationships on their own cognition because they have outgrown them. Of course, this is entirely voluntary. Those who desire to remain with a long-term mate can certainly do so. For those who wish to venture forth into new relationships, your guides will bring you more appropriate relationships since love (and sex which is a very sacred form of love when used as an expression of intimacy and creativity rather than lust) magnifies the recoder collective consciousness you are manifesting.

It is hoped that all Light Beings on Earth will elect to go through recoding. Yet, not everyone will complete recoding. Nonetheless, that possibility is extended to all. Bear in mind that whatever you attain will be sufficient for your evolution since it will fit with your soul contract. The speed with which you proceed through the nine segments of recoding will depend on how quickly you release your fears since fear will inhibit progress. However, recoding is a personal experience, and each of you must find your own pace. Those of you who move too quickly might damage your energy system which obviously defeats the purpose of recoding! As recoding progresses, the density of the third dimensional body will lighten and, in fact, you will begin to "glow." Ultimately, you will consist of more energy than density which will provide you with the ability to "remake" yourself into whatever you need at any point in time.

Those who are currently undergoing recoding are experiencing a variety of different sleep patterns depending on their reaction to the process. Some alter their typical sleep patterns, either moving from a long, hard sleep to a brief, light sleep or vice versa. Most will temporarily lose their ability to recall dreams. Since dreams are a window to your non-physical experiences, we block them to prevent you from the intense desire to remain with us. Although you cannot remember, as your astral body travels with us, it is a joyous experience as you reunite with old friends. Eventually, you will be able to recall your experience when we are assured you will not wish to leave Earth.

The Nine Segments

Although, as mentioned, recoding is not necessarily a sequential process, it will proceed in a similar manner to the following nine segments:

Segment One - Releasing Anger

Anger is an ongoing challenge during the recoding process because it has the potential of causing you discomfort. We ask you to release anger during this initial segment in order to eliminate negative energy prior to aligning and fusing your DNA strands, therefore minimizing your discomfort as the higher energy level of recoding enters the denser energy level of your present physical body. It is important that you are at peace with yourself as you expand in consciousness. Liver cleanses are great for moving quickly through segment one since the majority of anger is stored in the liver. In fact, any form of cleansing to release toxins is effective during this stage.

During this time, you should make peace with those who have angered you in the past. When working to release anger for past relationships, write a list of the names of those who have been involved and release them from your anger and blame. You may release them in person or at a soul level as long as you really feel the release of your anger toward them. Please note the use of the word "*release*" rather than "*forgive*" for this is an *extremely important differentiation*. Forgiveness is not part of Divine Creator truth because it is a value that assumes someone has done something wrong while the other has done something right. The Divine Creator holds no judgment so there can be no right or wrong. In the Creator's eyes, there is simply being, learning, and calibrating. Right and wrong are a result of the polarity on your planet, and it breeds judgment which separates beings. Separation keeps love from growing, thereby weakening you.

A more appropriate value than forgiveness is "*acceptance*" because it allows everyone to exist and learn at their own rate. Acceptance is a value you must embody to successfully recode since past efforts at helping third dimensional beings evolve to the fifth dimension have been thwarted by Light forces attempting to control the efforts of dark forces and vice versa. In other words, Light Beings directed energy at the dark forces instead of being solely concerned with their own progress. During this attempt to enable Earthlings to expand consciousness, everyone will follow their calling, whether dark or Light, because acceptance (implying no judgment) is the key to success. This also ensures no fear since fear is a negative frequency which will undermine a recoder's success. Ideally, as the Light beings grow in strength, they will draw the others toward the Light through example.

Segment Two - Managing Anger

If you stuff anger during recoding, you may place yourself in pain. Segment two ensures that anger does not rebuild in your system after clearing it in segment one, thus re-creating the density from which you are attempting to separate. It also protects you and others since segment six in the recoding process gives you the ability to manifest your thoughts very quickly, both negative and positive, an attribute that could cause more harm than good unless you have moved through segment two. Stuffing anger may cause discomfort when recoding since it recreates the toxins that have been eliminated. Do not stuff anything when facing conflicts. As you go through recoding, you will find that you will feel more neutrality toward situations rather than a completely irrational surge of anger. You will also experience greater compassion. However, do not expect to be able to completely remove anger from your repertoire. After all, you reside in a human body, and anger is an aspect of your experience. You should expect to respond with the same emotions you have always had.

In the case of anger, you may vent in a vexed voice at a particular situation. But, you should no longer stuff the anger because you have not communicated how you have felt about a particular situation. Experience the release of your anger as you emote. Don't harbor anger by fixating on it. It is a "momentary" release, not a lifelong commitment to how you felt at that moment.

Segment Three - Clairaudience

Your telepathic channel begins to open in segment three. As you clear the lower emotions from your system, you become lighter. Your guides are better able to communicate with you, either through channeling, automatic writing, or whichever telepathic form you wish to use. This is the point at which you can speak directly to your guides rather than getting the information via your dreams, intuition, or through another individual who has psychic abilities. If you resist what you begin to hear as your channel begins to open, there may be discomfort in either your right or left ear or in your third eye (central brow) which is located about an inch above the eyebrows midway in the forehead. There may even be some blockage or earaches, but do not be alarmed since it means the channel is opening and you need to support that transition. It is important to avoid heavy electrical environments where the current flows in a random state because it causes static in your channel and may result in dizziness or headaches.

Electricity flows randomly in locations like amusement parks, casinos, heavy computer areas like telemarketing centers, electrical plants, nuclear plants, etc. Less static occurs in heavy electrical environments like airplanes or office buildings since the electricity moves in a single direction. Interference from electricity will persist from segments three through five as the clairvoyant channel continues to open. Remove electrical devices from your sleeping environment, particularly electric blankets, heated water beds, and TV's. If you sleep near too many electrical devices, it is difficult for your astral body to remain clear. It is recommended that you lessen your TV-watching and radio-listening since these frequencies carry some negative energy.

Although certain types of music are healing, the appliances that play the music are detrimental due to the electrical currents. If you must listen to music, be sure that the electrical device is not in direct alignment to your heart chakra. At this time, your own personal channels are developing so spend time with them instead of with the negative content from the air waves. Get back to nature because this enables your genetic engineer to make more progress since there is no interference in the form of static. During segment three (and four), work begins on your astral body during sleep. Your astral body visits with your genetic engineer who works on realigning and fusing your twelve strands. During segment three, you will be unable to remember these visits as you will want to remain rather than return to the more restrictive third dimension. Later, you may recall your experiences when you are able to use enhanced abilities on Earth and have less resistance returning.

You may experience some discomfort as your genetic engineer "rearranges" your DNA although we will do our best to minimize it. Efforts on your part to stay cleansed via therapies like Reiki, Holographic Repatterning, acupuncture, massage, chiropractic therapy, liver and colon cleanses, healthy diet, and a host of other options will reduce your potential discomfort. When your astral body is away, a "caretaker" will be left with you so no one will know you are elsewhere. The caretaker is a physical and mental copy of you. However, this assisting soul will need to be trained by you in order to do a good job since it is not an emotional or spiritual carbon copy of you. If you wish your caretaker to replicate you as best they can, you must give him or her permission to read your soul records or tapes. These are private and other beings are not allowed to use them without permission. More information is provided about training your caretaker in segment five.

Segment Four - Clairvoyance

Segment four will open your third eye so you will be able to "see" your guides and any other entities. This is a gradual process since your third eye has atrophied through lack of use in third dimension. Clairvoyance may begin with the sensing of dark forms followed by lighter forms and, finally, a full color range. Some people see the forms in their everyday physical settings and others see them psychically. Most of you will see flashes of light or dark out of the corner of your eye as well since it is easier to use peripheral rather than direct vision to catch a glimpse of a non-physical entity. When your third eye opens and if you begin seeing fourth dimensional entities next to friends and loved ones, there may be a tendency for you to experience fear. However, fear will keep you from moving through this segment so overcome it as quickly as possible. Most of the time, fourth dimensional entities who are dark energy will not harm you since they are more interested in the being they are stalking. However, should you feel uncomfortable in their presence, shine your Light brightly around yourself as well as the person you are with and they will retreat during the time you are present.

Segment Five - Integration

Segment five is critical because it builds the foundation for the final four segments. It is a segment that some will elect not to clear due to the fact that it requires acceptance of your soul contract which may require some alterations to your life. These alterations will seem like natural choices to you if you are interested in continuing to grow, i.e., leaving unfulfilling relationships behind, changing careers, shifting behavior patterns. This integration provides the focus you will need for your work during the final stages of recoding. In order to create your two-stranded vehicle, implants were inserted in your astral body to prevent ten of the twelve strands of DNA from magnetizing together in a twelve strand helix. The twelve strands are scattered throughout your energy body and, as each implant is removed, they begin to return to their original configuration. Once they are in place, a fusion must occur to reconnect the strands. If we realigned and reconnected all twelve strands at the same time, it would cause energy damage in your auric field which would not achieve your objective! Therefore, we must proceed in a delicate and cautious manner, taking each step when it is most beneficial to you.

There may be some things that will not clear in segments one through four due to the method in which we must reconnect the twelve strands. It will be necessary during this time to go back and ensure that everything has been integrated. Also, it is sometimes not possible to fully complete each segment during each session because it may cause stress on your physical body. Your genetic engineers need to wait to complete the task when your body is ready. During segments five through seven, your caretakers will be left for longer intervals since we will need your astral body for up to five days at a time.

You may experience some discomfort when the caretakers reside for longer periods inside your physical body since they have never lived in the third dimension and do not have an affinity for negative energy. If you are still carrying negative energy, they will resist it by pushing against it which may make it feel worse. This is a time to increase your energy cleansing routines like Reiki, acupuncture, massage, colonics, and so forth. It is also imperative that you make a "to do" list for your caretaker in order for them to be productive. For example, tell your caretaker that they must arrive at work on time every day, that you always accomplish your job responsibilities, that you routinely exercise, or that you are on a diet and do not eat sweets. Train them to be more of a "housekeeper" than a "housesitter." If you are dissatisfied with how your caretaker manages life while you are away, request a new one. Caretakers are young souls who wish to experience life in the third dimension, and they are novices at it. This is a growth experience for them, too. If your caretaker prefers to lounge in bed all day rather than clean your house, you have the right to request one who complies with your wishes. Additionally, caretakers have no heart connection to your life since they are copies and do not carry emotion. Do not expect your caretaker to experience the highs and lows that you take for granted as part of your third dimensional life. Caretakers will be more neutral and passive, something that may be remarked upon by your friends and family.

Segment five is exciting because you can begin to transfer to other dimensions, simultaneously experiencing the multiple lifetimes you are playing at the same time that you are on Earth. During this time, bleedthrough might occur which may come through in the form of dreams or feelings of deja vu. However, you will know that it is not a dream because it is very realistic and sequential. Segment five is also a time when you must focus on your soul contract. As we stated, some will elect not to move past this segment because of the commitment required. However, for those who wish to proceed, we must be assured that you are committed to your soul contract before reuniting you with your lost power.

Segment Six - Owning Your Power

By the time you reach segment six, your guides are ninety-eight percent positive that you are firmly on your path and you will not back off your soul contract, so you begin to retrieve greater levels of power. It is the segment that enables your mind to manifest whatever it desires quickly, bringing you what you desire (or not desire if you are not careful). This is a time to keep clean, positive thoughts since negative thoughts are manifested equally quickly.

Segment six is tricky since you are no longer relying on your guides to manifest for you. You are finally doing it yourself. You have been taught to ask your guides for assistance and to state affirmations in present tense to realize what you want. However, do not be surprised if your guides gently remind you to do it yourself once you reach segment six since they want you to recapture your power. For example, if you desire financial abundance, experience the decisions and resulting emotions that come with unlimited wealth.

If you wish to have your life mate join you at this time, view yourself as a radio frequency that can broadcast a message and send a message full of love to your future mate. Support that signal with the feelings you will have while with your life mate, experiencing the emotions of sharing time or making love with your loved one. The more intense the feelings that you generate during this time, the faster you will bring what you desire (as long as you are not holding conflicting feelings that discreate as fast as you create).

For those of you who have read *The Tales of Alvin Maker*, you have arrived at the point where *you are the maker* rather than a passive receiver of what your guides can do for you. Your guides play the role of assister rather than maker since that is your role. This is a transition from the "help me" requests of your spirit guides to the recognition that you own your own power and need to begin flexing your manifestation muscles. Bear in mind, at this point less than half of your strands of DNA are connected which limits your ability to move to total manifestation of your wishes. Use this time to increase your manifestation strength in preparation for segment eight which will provide you with more power.

Segment Seven - Removing Illusions

Segment seven can be described as follows: you feel the presence of a light at the end of a long dark tunnel which symbolizes your mission, and you must traverse the tunnel to reach the light. Your personal power is the torch that you light to see as you move through the darkness toward your mission. Every time you hit an obstacle, you use your torch (or power) to shed light on the obstacle, thereby exposing the illusion that has been created. In segment six, you learned how to be a maker. Now, in segment seven, you have become a creator. By the time you experience segment seven, you will probably have approximately eight strands of DNA aligned and fused which will assist you in your creation efforts. Creators can manifest any reality they choose since they realize the "physical" world is based on illusion. Whenever you hit a perceived obstacle but choose to identify it as an illusion, you shift to a positive creator. Unfortunately, you also act as a creator when you hit a perceived obstacle and choose to identify it as a problem. As the creator, you can create any reality so remember to choose the positive rather than negative path.

Segment seven is difficult because you move from the realization of yourself as the power source in segment six to running the gauntlet to test that power. You will also probably begin to attract more attention from sources of dark energy. They are mesmerized by this type of light energy since it is ultra-powerful and would be a tremendous boost to their own energy if they can share or capture it. If you do not already have a daily routine of placing protection around you, now is the time to start. Place a shield of protection around you twice daily, once upon rising and once at night. Use whatever protection method works best for you. Or, envision gold light in the form of a mesh. Encircle yourself with the mesh and zip it up just like Cinderella was secure in her pumpkin. Request additional protection from your guides, genetic engineer, and caretaker as well since Spirit cannot interfere in a free will zone without permission.

Segment Eight - Releasing Fear

By segment eight, either all or most of your DNA strands will be fused and activated so much refinement will occur in how the brain handles the additional strands. This work proceeds quickly and your astral body will not need to spend extended time periods with the genetic engineers. Crystals that contain your codes, just like a computer chip, are accessed. Each crystal is reprogrammed which enables the DNA to carry additional information throughout your body. The firings or synapses of energy sequences throughout the brain will operate differently once your implants have been reconditioned, especially since the sequences are firing into additional DNA strands.

Also, by the end of segment eight, all of your fear implants will be removed. These implants were placed in your body prior to entering Earth in order to allow you to maximize your experience of third dimension which holds fear as a prevalent emotion. Fear is a tremendous learning tool which allows you to travel in the darkness, recognize it for what it is worth, and determine whether or not you choose to return to the Light. This may sound masochistic to you, but these implants serve as a protective device since many of you came from worlds that held no fear. You would have been completely vulnerable on Earth without fear which is a handy emotion to have when living on a planet filled with violence. After becoming street-smart, you can lose that fear and still survive. In fact, you excel at survival because you bring a more developed consciousness of life without fear to the planet.

Whereas segment six was a time to take back your power and begin feeling the importance of your role as a maker, segment eight provides you with the juice to immediately manifest. This is when you reap the rewards of learning how to harness your own energy rather than being on automatic pilot with your guides managing the manifestation energy. Segment eight clears the illusions from segment seven and allows you to completely manifest whatever you choose as long as you use your own power.

Segment Nine - Freedom from Guilt

This segment is the crème-de-la-crème of the recoding experience. You now have your twelve strands of DNA aligned and fused. It should feel like pure pleasure to you since all of your guilt implants are now gone along with your fear implants. This is the first time you have experienced life on Earth without fear and guilt implants. Imagine the full capacity of your energy without fear and guilt! Enjoy every minute of this experience since you are one of the few beings living on Earth who no longer holds fear and guilt in your mental body.

Although you still reside in a dense body, your energy has lightened significantly since you are vibrating at a higher frequency. Your frequency will continue to lighten as the Earth increases its own frequency, enabling you to move energy more quickly and easily. Energetically, spend time exploring the twelve levels of awareness to which you now have access. Remember, you are privy to an expanded experience based on your commitment to recoding.

Summary

In the beginning, you may be frustrated since you can view the total picture yet your energy is still contained in a three dimensional body. However, this frustration will dissipate with time as the energy continues to lighten on Earth, allowing you to experience more and more benefits from your twelve strands. Expect for others to notice you since you will project an inner and outer glow from your increased vibratory frequency. You will be the life of the party and will easily attract those who need to be with you or work with you.

Appendix B

- Phase II -

Plug-in & Endocrine System Balancing

Plug-in Process

The following information was transmitted from Joysia, Chief Genetic Engineer, during late December, 1996 and early January, 1997. It includes the information needed to connect the twelve DNA strands to the physical body through the endocrine system. The endocrine system is in the process of replacing the immune system as the primary system in the human body. The immune system is a defense system which operates on the premise of fighting foreign material that enters the body. However, one cannot be balanced while undergoing resistance. The endocrine system is a support system that operates on the principle of balance. It is the path to Divinity. The endocrine system must replace resistance with balance or Divine Union in order for humans to move to the next stage of evolution.

After you have completed your work in the nine recoding segments and have connected your twelve strands of DNA *energetically*, it is necessary to connect the twelve strands into your endocrine system. This is called the "Plug-in" Process. Plug-in occurs after your completion of the nine segments because it requires a higher (or lighter) energy vibration to be present in the body. Bringing the twelve strands of DNA into your energy field significantly increases the rate at which you vibrate. Think of it like a mainframe computer compared to a personal computer. The mainframe computer uses more power, and it can hold more data and process faster versus the PC. You have evolved from a PC to a mainframe capacity which enables you to run more data through your system.

The power from the additional DNA strands is run into your endocrine system which is a "liquid" system because the liquid acts as an insulator for the additional power. There are also some sacred geometry principles present in the liquid that enable us to use molecules with pyramidal structures to translate the incoming energy into a language that your body will understand. This pyramidal structure helps to "hold" the energy as well because that particular molecular shape is receiving an energetic "hook up" from a more highly evolved frequency, one that is not yet present on your planet. Essentially, you are grounding that energy on the planet through your agreement to complete this process. Once critical mass is reached when enough of you carry the frequency, those who are born during or following the attainment of critical mass will be able to carry the twelve strands physically in their body.

As you progressed through the nine segments of recoding, you performed a general emotional housecleaning. You removed old blockages and negative energy from your system, becoming as emotionally balanced as possible by removing the negative, denser energy in order to reduce the discomfort you might have experienced from a high energy frequency meeting the resistance of the dense energy. Upon completing the nine segments, you are free of much of your "old stuff," minus some residual negative energy that exists in the cell structure of your physical vehicle that takes longer to clear. However, there is still some tweaking that needs to occur to bring your endocrine system into perfect balance for Plug-in.

The tables on pages 184 & 185 (12 Strand and Endocrine System Balance Sheet) have been developed to help you focus on the remaining areas needing balance. Once balance occurs, we can complete the Plug-in process over three days and three nights. This is conducted remotely. It does not require the presence of your astral body which eliminates the need for you to experience a caretaker in your body as you did during recoding.

The chart shows the twelve strands of DNA across the top. Each strand relates to a specific emotional dimension. For example, strand one represents Courage. The outlying emotions that cause the optimal state to move out of balance are listed below each emotional dimension. In the case of courage, a person is out of balance when they are too courageous (i.e., Warrior) as well as when they are not courageous enough (i.e., Victim). The endocrine glands involved in the Plug-in are listed on the left side of the chart. Each of the endocrine areas relate to a life stage. For example, the hypothalamus corresponds to pre-birth, that time prior to entering the fetal body which was used in preparation for this lifetime.

It is recommended that you dowse using a pendulum or muscle test using kinesiology to determine the boxes in which you have already achieved balance, asking the question, "On the (ENDOCRINE GLAND), is the (NUMBER STRAND) balanced?" For example, "On the hypothalamus, is the number one strand balanced?" If you do not know how to use a pendulum or muscle test, contact someone who does. Continue asking this question for each DNA strand and for each endocrine gland until you have dowsed each cell on the Endocrine Balance Sheet. Circle each strand that is balanced.

Your chart will contain a series of circles that will create some interesting patterns. Looking across the page horizontally, you will notice that there are entire life stages (endocrine system related) that contain no check marks. Looking down the page vertically, you will notice entire emotional dimensions that are blank. This gives you some clues as to what you need to bring into balance for Plug-in. However, it is not a total picture, since much of what you will need to clear comes from soul commitments or vows made during prior lifetimes. Although you did not experience the situation in this lifetime, you carry an imbalance for a particular life stage or emotional dimension based on a prior situation.

The following process will bring all of the boxes into balance in order for us to proceed with Plug-in. They should be conducted in the following order. After you have completed each step, dowse your chart to determine what has been brought into balance.

1. Set the intent to have everything in balance within certain dates. Ask the Divine Creator and your guides and teachers to help you in this endeavor.

2. Find out if you need a soul-clearing by asking your spirit guides, dowsing with a pendulum, or muscle checking through kinesiology, asking "Is it necessary to have a soul-clearing in order to successfully proceed with the Plug-in Process at this time?" If you do need a soul-clearing, determine which one of the recommended practitioners listed below is best suited to do a soul-clearing for you. Or, if you know someone who performs this type of work, ask them to assist you. A soul-clearing not only clears the Akashic records of any past life experiences that are causing you to carry

negative energy at a soul level, but it also ensures that you can
ultimately return to your home planet or star. You have picked up
negative karma in your Earthly reincarnation cycles, as you intended, in order to experience the polarity of light and dark.
However, you cannot bring remnants of your dark ties into a light
society. It would be like bringing measles to Hawaii, polluting an
environment that does not contain this type of negativity. Some
recommended soul-clearing practitioners include:

- Ann Brewer, 913-722-5498
- Venessa Ralston, 520-204-0896

3. At this point, you will need to have completed the work on the
nine segments of recoding. Check to ensure you have completed
and cleared all nine segments. If you have not, determine where
your blocks are and complete and clear those segments. *Unfortunately, we cannot proceed with Plug-in unless you have finished
recoding.*

4. Open yourself at least 95% to all Archangel realms. There is an
exercise on page 177 which will assist you in this process. If you
run into blocks and are unable to achieve 95% in a particular realm,
ask a soul clearing or energy practitioner for assistance.

5. Rescind all past life vows that are blocking emotional balance.
There is a sheet on pages 178-179 for Level I and Level II of this
process.

- Level 1 is a "generic" vow-clearing that covers commitments
  to conditions like poverty or chastity, i.e., vows possibly
  made for your higher good as a priest or priestess in past
  lifetimes that no longer serve your higher good. Level I also
  clears vows of obedience that were made to beings
  perceived as Divine Creators (e.g., Mayan priests who
  performed ritualistic rape and sacrifice for bountiful harvests)
  that are obviously no longer pertinent to your higher good.

- Level II clears vows that you made based on misperceptions
  of prior life experiences. For example, you may have desired
  to have a life experience as a healer but were burned at the
  stake for witchcraft, crying out during your death that you
  would "never speak your truth again." This vow of

"protection" no longer serves a purpose and may even interfere with your soul growth.

Please note that it may also be necessary to conduct some past life regression work with a trained hypnotherapist or psychic counselor if you have difficulty calling up your past lives through these exercises.

6. At this point, you will probably want to go back to the 12 Strand and Endocrine System Balance Sheet and determine if any other strands have balanced as a result of your soul work via the soul-clearing, Archangel realm work, and disavowal of vows. Usually, more strands are brought into balance by these processes.

7. Follow the Balancing Modalities listed for each strand that is not circled on your 12 Strand and Endocrine System Balance Sheet. You do not need to do the healing modalities if a box has already balanced. Please note that this is not a linear procedure. For example, you may need to complete a series of Balancing Modalities in a single column prior to achieving balance in a single box, i.e., one box's balance is contingent upon another box's balance. Or, you may clear several in one column and find that others in the same or a different column also become balanced. Or, you may encounter a life experience or lesson, and the way in which you address this situation will balance a box. This means you will need to reconfirm which of your boxes are balanced while you are working on this matrix since you may balance one without being aware you have done so. Also, bear in mind that some boxes may not clear at all, requiring a matrix of boxes to be checked across the page and the three-day/three-night Plug-in process to be completed prior to balancing them. Balance as many boxes as you can through the Balancing Modalities, dowsing to determine balance. Try to have *no more than six boxes empty* when requesting Plug-in.

8. When almost all of your boxes are checked, you may request Plug-in. Ask your genetic engineer "Is this 12 Strand and Endocrine System Balance Sheet balanced enough for Plug-in to take place?" To initiate the Plug-in, light 12 white candles and place them around your bath tub. As you light the candles, say:

- "Divine Creator (3 times), My Guides and Teachers (3 times), All Seven Archangels Loyal to the Divine Creator (3 times). Please come forward and assist me in balancing my energy field for Plug-in to commence. Bless these candles as beacons of energy that represent my 12 DNA strands, connecting the energy to my endocrine system in order for me to meet my soul destiny contract of carrying the twelve-stranded energy on Earth during its time of transition. I lay myself before you in this water to symbolize the liquidity of my endocrine system, asking you to proceed with this sacred ceremony."

Lie down in a salt water and baking soda (1 cup each) bath. Visualize the following colors to balance your chakras at each of the chakra locations:

- Red: root chakra (base of spine)
- Orange: pelvic chakra
- Yellow: solar plexus chakra
- Green: heart chakra
- Blue: throat chakra
- Indigo: brow chakra
- Violet: crown chakra

Then, envelope yourself in white light. Lie very still. You will feel the energy entering your body. When you stop feeling the energy, you can get up. The Plug-in will be completed during the three days following that initiation point. If, for any reason, you begin to experience discomfort, contact a healing practitioner for assistance.

Releasing Blockages to Archangel Realms*

Percent Open

Archangel Kamiel
Light, Power, Energy   _____

Archangel Zadkiel
Materialization, Creative Visualization   _____

Archangel Zophkiel
Art, Beauty, Perfection, Balance   _____

Archangel Auriel
Nature, Compassion   _____

Archangel Gabriel
Sound, Vibration, Communication   _____

Archangel Michael
Protection, Truth   _____

Archangel Raphael
Healing, Love   _____

The seven Archangels provide you with a specific energy, teaching, and aspect of the Creator. However, sometimes your past/present life issues, traumas, or subconscious programming may have created blockages between you and these seven Divine Realms. You need to remove those blockages.

While dowsing with a pendulum, ask, "What percentage am I open to the realm of Archangel _____?" Are you more than fifty percent, more than sixty percent etc., until you have determined the percentage. You will discover that you can be one hundred percent in some realms and blocked in others.

To release blockages, light a white candle for each Archangel Realm with which you wish to be 100 percent open. As you light each candle, speak the Archangel's name three times and request a full release of your blockage and a Divine reconnection with the energy and teaching of that specific realm. Allow the candles to burn as you experience the energies shift within you. When you feel complete, blow out each candle and close the ceremony with three "Thank you's."

*Releasing Blockages provided by Venessa Ralston

Disavowing Vows*

Level I

Commitments to Conditions

- Light a white candle. Request the Divine Creator's presence by repeating Creator of All Beings three times each. Request the Sirian/Pleiadian Council by repeating Sirian/Pleiadian Council three times each. Request the Akashic Record Division by repeating Akashic Record Division three times each: "Creator of All Beings (3 times), Sirian/Pleiadian Council (3 times), Akashic Record Division (3 times)"
- Connect with Creator energy by tapping on your thymus gland (located approximately 2 inches below base of neck at top of chest) as you intone these syllables three times each: "Ahh (3 times), Ooo (3 times), Umm (3 times)"
- Say: "I wish to disavow all vows I have ever made in ignorance or in truth to any God or perceived God, any brotherhood or sisterhood, any other entity, or to You as the Creator, that are no longer working toward my higher good. I particularly wish to disavow all vows I have ever made that are blocking my ability to balance and reconnect my twelve strands of DNA into my endocrine system. Please assist me in disavowing these vows so that they are no longer a restricting device on my being. I ask that all restrictions be lifted and eliminated so that I can expand and align with my higher path and purpose. In the highest good of all, thank you, thank you, thank you."
- Blow out the white candle and say "So let it be."

*Level I exercise provided by Venessa Ralston

Disavowing Vows

Level II

Based on Misperceptions of Past Life Experiences

- Darken the room.
- Light two white candles and place one on either side of you.
- Look directly into your eyes in a mirror for fifteen minutes. (Set a kitchen timer if you need to.)
- Say "I, of my own free will, wish to bring to the surface all lifetimes in which I have made vows that now need to be rescinded. So be it."
- Sit patiently and allow your mind to bring forward visions of past lives in which you made vows that no longer serve your highest good. As you view these situations, thank yourself for the protection these vows gave you in the past. Release the present need for them.
- Repeat this three days in a row.

Balancing Modalities

The following ninety-six modalities have been developed to assist you in balancing your endocrine system in order to undergo the Plug-in process. The endocrine system has to be balanced in order to connect the twelve strand energy frequency without causing stress on you physically. Please note that the following items must also be completed prior to using the balancing modalities since the modalities will not override and clear any incomplete work from these steps: 1) the nine segments from phase one of Recoding; 2) a soul clearing or elimination of negative energy from past/present lifetimes residing at a soul level (check to see if this is necessary by dowsing or asking an intuitive counselor); 3) elimination of blockages in the seven Archangel realms; 4) disavowal of vows from Level I and Level II. These items have been discussed in detail on pages 151-179.

At this point, you should have dowsed your 12 Strand and Endocrine System Balance Sheet to determine which strands are balanced. It is now time to utilize the balancing modalities for any remaining unbalanced areas. Sometimes, you only need to conduct the balancing modality on one endocrine gland of one strand for the rest of the column to clear since there is a domino effect. Sometimes, you need to balance the entire column. You do not need to proceed in any particular order. Pick and choose the boxes that you most wish to balance. Reference your 12 Strand and Endocrine System Balance Sheet. Let us say, for example, that strand three (Creation) is not balanced for the thyroid gland (The Soul). This is reflected by the absence of a circle around the number 3.4 in the box in column three, row four. Turn to page 196 for the balancing modalities on strand three. Locate 3.4: "The Soul" and complete the balancing modality described for it. After you have completed the modality, your box will be balanced when you ask the question, "On the thyroid gland, is the number three strand (3.4) balanced?"

Each balancing modality contains the following information: a color that corresponds to the endocrine gland, the meridian affected by the lack of balance, an affirmation that represents the balanced state in physical form, and the modality for bringing that condition into balance. Please note that three charts have been provided in addition to a verbal description of the location of the meridian to assist you in finding it. Sometimes, you will need to utilize the color, the meridian, and the affirmation to bring the strand into balance. Sometimes you will only need to use one of the three items.

The balancing modalities have specifically been developed to be as simple as possible in order for everyone to have access to the materials needed. Frequently, you will use color and breath to clear. Sometimes, you will need to purchase a candle or a pitch pipe (if you do not own a musical instrument) to complete the work. Do not be fooled by the simplicity of the modalities. Many people have used them, and they work!

Remember, you can conduct the Plug-in ritual described on pages 175-176 as soon as *no more than* six boxes are left unbalanced. The remainder can be balanced without any stress to your physical body.

Twelve Strand and Endocrine System Balance Sheet and Balancing Modalities

PLUG-IN PROCESS: 12 Strand and Endocrine System Balance Sheet (Strands 1-6)

	Strand 1 **Courage** Warrior/Victim	Strand 2 **Concentration** Compulsive/Inattentive	Strand 3 **Creation** Greedy/Self-sacrificing	Strand 4 **Well-being** Strong/Weak	Strand 5 **Joy** Hedonist/Martyr	Strand 6 **Truth** Manipulating/Pleasing
Hypothalamus Pre-birth	1.1	2.1	3.1	4.1	5.1	6.1
Pineal Soul Entry	1.2	2.2	3.2	4.2	5.2	6.2
Adrenals Instinct	1.3	2.3	3.3	4.3	5.3	6.3
Thyroid The Soul	1.4	2.4	3.4	4.4	5.4	6.4
Gonads Youth	1.5	2.5	3.5	4.5	5.5	6.5
Thymus Independence/Interdep.	1.6	2.6	3.6	4.6	5.6	6.6
Pituitary Growth	1.7	2.7	3.7	4.7	5.7	6.7
Heart Maturity	1.M	2.M	3.M	4.M	5.M	6.M

PLUG-IN PROCESS: 12 Strand and Endocrine System Balance Sheet (Strands 7-12)

	Strand 7 Balance Over/Under	Strand 8 Power Controlling/Enslaved	Strand 9 Communion Solitary/Communal	Strand 10 Cohesiveness Independent/Dependent	Strand 11 Sexuality Male/Female	Strand 12 Love Aggressive/Servile
Hypothalamus Pre-birth	7.1	8.1	9.1	10.1	11.1	12.1
Pineal Soul	7.2	8.2	9.2	10.2	11.2	12.2
Adrenals Instinct	7.3	8.3	9.3	10.3	11.3	12.3
Thyroid The Soul	7.4	8.4	9.4	10.4	11.4	12.4
Gonads Youth	7.5	8.5	9.5	10.5	11.5	12.5
Thymus Independence/Interdep.	7.6	8.6	9.6	10.6	11.6	12.6
Pituitary Growth	7.7	8.7	9.7	10.7	11.7	12.7
Heart Maturity	7.M	8.M	9.M	10.M	11.M	12.M

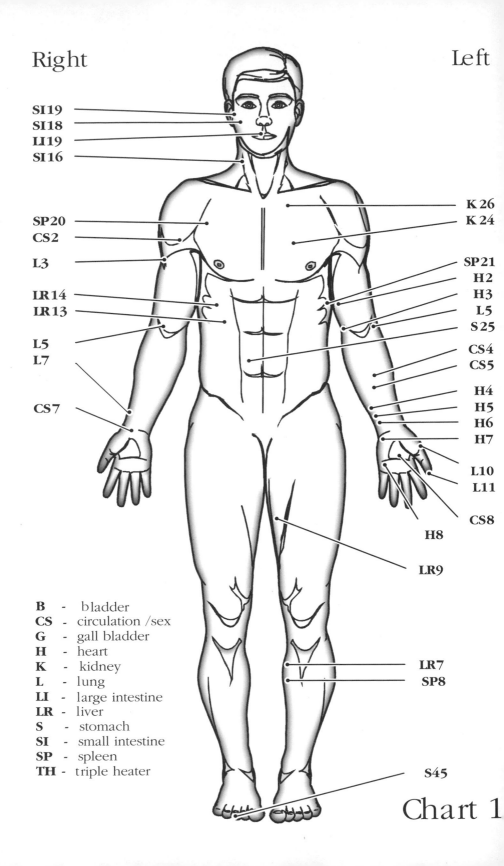

Right

Left

SI19
SI18
LI19
SI16

K 26
K 24

SP20
CS2
L3

SP21
H2
H3
L5
S 25

IR14
IR13

CS4
CS5

L5
L7

H4
H5
H6
H7

CS7

L10
L11

CS8

H8

LR9

B - bladder
CS - circulation /sex
G - gall bladder
H - heart
K - kidney
L - lung
LI - large intestine
LR - liver
S - stomach
SI - small intestine
SP - spleen
TH - triple heater

LR7
SP8

S45

Chart 1

Left

Right

G20

SI14
SI12

TH12

B45
B22
SI8

TH8

TH2

B54

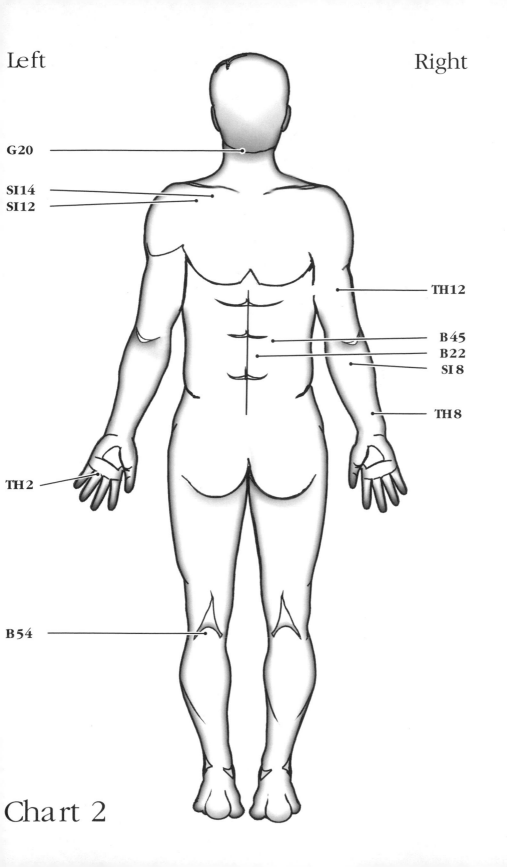

Chart 2

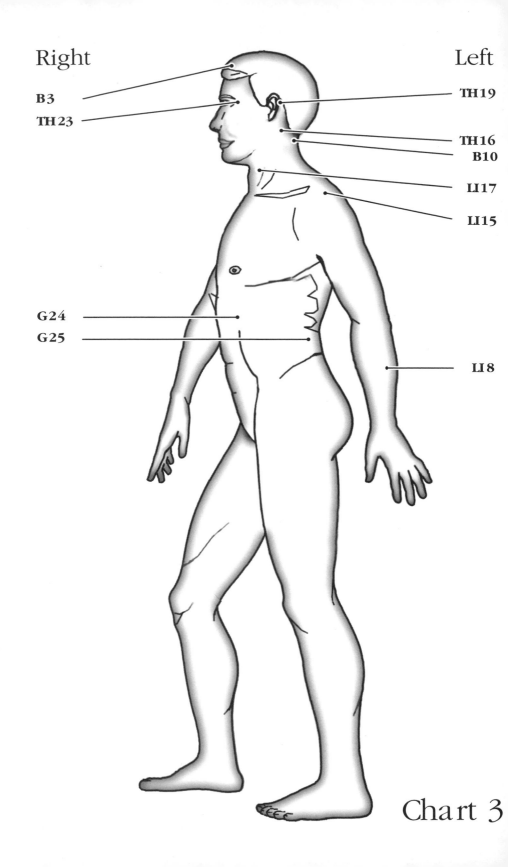

Right

Left

B3

TH23

TH19

TH16
B10

LI17

LI15

G24

G25

LI8

Chart 3

Strand 1 - Courage

1.1: Pre-birth
Color: Magenta
Meridian: Small intestines, point #16, located behind the right SCM muscle (on side of neck) on level with the Adam's apple (Chart 1)
Affirmation: I am courteous.
Balancing Modality: Suck air quickly in through your mouth and hold it while you pump your abdominal muscles out seven times. Exhale with 7 breaths through your mouth, pushing out the abdominal muscle with each exhalation. Visualize the color magenta being inhaled into your body on the inhalation breath. Repeat this sequence 7 times. Then, hold your right hand over your right point #16 small intestines meridian and say 7 times aloud, "I am courteous."

1.2: Soul Entry
Color: Violet
Meridian: Heart, point #7, located on pinkie side of left hand at point where the hand meets the wrist. (Chart 1)
Affirmation: I delegate according to an individual's best talents and gifts.
Balancing Modality: Inhale through your nose and exhale with your tongue touching the roof of your mouth, making a long "SSSSS" sound like a snake. Repeat this 20 times. Then, exhale all of your air and hold your breath, relax deeply, and imagine you are breathing internally for as long as you comfortably can. Hold your right hand over your left point #7 heart meridian and say 2 times aloud, "I delegate according to an individual's best talents and gifts." Visualize the color violet glowing and expanding in the left point #7 heart meridian until you feel it has "taken hold."

1.3: Instinct
Color: Orange
Meridian: Heart, point #7, located on pinkie side of left hand at point where the hand meets the wrist. (Chart 1)
Affirmation: I know my purpose in life which supports my "true" way.
Balancing Modality: Let your jaw hang down loosely with your tongue resting on the floor of your mouth, and gently pant in and out through your mouth like a dog. Your body should feel loose and relaxed. Breath this way for about 2 minutes. Say 4 times aloud, "I know my purpose in life which supports my 'true' way." Visualize the color orange glowing and expanding in the left point #7 heart meridian until you feel it has "taken hold."

1.4: The Soul

Color: Blue

Meridian: Gall bladder, point #24, located approximately 5" below left nipple in line with the nipple. (Chart 3)

Affirmation: I embody integrity in every thought that I have.

Balancing Modality: Place your right thumb on the side of your right nostril to block the air flow. Exhale and inhale through your left nostril. Then, place your right forefinger on the left nostril. Exhale and inhale through your right nostril. Alternate nostrils 4 times, completing 2 cycles of breathing. Place your left hand on the left point #24 gall bladder meridian and say 4 times aloud, "I embody integrity in every thought that I have." Visualize the color blue and bring it into your heart, feeling the color merge with the feeling of peace you have in your heart when you live in integrity.

1.5: Youth

Color: Red

Meridian: Small intestines, point #16, located behind the right SCM muscle (on side of neck) on level with the Adam's apple. (Chart 1)

Affirmation: I am courteous.

Balancing Modality: Place your right thumb on the side of your right nostril to block the air flow. Exhale and inhale through your left nostril. Then, place your right forefinger on the left nostril. Exhale and inhale through your right nostril. Alternate nostrils 4 times, completing 2 cycles of breathing. Visualize the color red expanding through your entire body, entering through your crown chakra. After you have established this color throughout your body, say 2 times aloud, "I am courteous." See the color red flowing from your body at the right point #16 small intestines meridian, letting it in through the crown chakra and out through the right point #16.

1.6: Independence/Interdependence

Color: Green

Meridian: Liver, point #14, located about 4 inches under right nipple in line with right nipple. (Chart 1)

Affirmation: I envision all possibilities.

Balancing Modality: Inhale through your nose. Exhale through your mouth. Now wait for the body to determine when it needs to repeat the inhale. While waiting, relax and let go, repeating the affirmation "I envision all possibilities." Visualize the color green exiting your body at the location of the right point #14 liver meridian as you exhale. Repeat this sequence 3 times.

1.7: Growth

Color: Indigo
Meridian: Gall bladder, point #20, located at the back of the neck where the hairline meets the spine. (Chart 2)
Affirmation: "I move forward with purpose and make the necessary changes that are needed," and "I have clear vision as I move forward."
Balancing Modality: Say 2 times aloud, "I move forward with purpose and make the necessary changes that are needed." Touch the color indigo (use a swatch of fabric or a candle) as you are stating the first affirmation. Say 2 times aloud, "I have clear vision as I move forward" while covering the point #20 gall bladder meridian with your left hand.

1.M: Maturity

Color: White
Meridian: Heart, point #7, located on pinkie side of left hand at point where the hand meets the wrist. (Chart 1)
Affirmation: I am open to every experience.
Balancing Modality: Hold your left hand over the point #7 heart meridian. Visualize an earlier experience when you were not open to what you were receiving from that experience. Envision a different ending, one that would have occurred had you been in flow with the process. See the entire setting bathed in the white light of Divine Creator energy, feeling the purposefulness of the experience and how, when you are open, the entire situation aligns with your true purpose. Lift your left hand up to the heavens and feel the freedom that surges.

Strand 2 - Concentration

2.1: Pre-birth
Color: Magenta
Meridian: Lung, point #3, located on the right biceps even with the nipple line. (Chart 1)
Affirmation: I take in all that is wise and inspiring.
Balancing Modality: Light a magenta candle. Say 2 times aloud, "I take in all that is wise and inspiring." Breathe in and out through your nose, drawing in the color magenta from the candle as it burns. When you exhale, radiate the color magenta around your entire body, lighting up your body. Repeat this breath 2 times. Hold your left palm over the right point #3 lung meridian using your left hand. Keep the meridian covered for 1 minute. After completing this balancing modality, let the candle burn without extinguishing it.

2.2: Soul Entry
Color: Violet
Meridian: Large intestines, point #8, located on the left forearm, just below the elbow on the outside of the arm. (Chart 3)
Affirmation: I accept myself.
Balancing Modality: Inhale deeply through your nose and exhale with your tongue against the roof of your mouth, making a long "SSSSS" sound like a snake. Repeat this 3 times. Next, balance your point #8 large intestines meridian by doing the following meridian massage. Lightly stroke up the outside of your right arm with your left hand 3 times, starting at the tips of your fingers and brushing up over your shoulders and up your neck and the side of your face. Repeat this using your right hand and brushing up your left arm. Say 4 times aloud, "I accept myself."

2.3: Instinct
Color: Orange
Meridian: Heart, point # 3, located on the inside of the left arm, just above the elbow. (Chart 1)
Affirmation: I am impartial, and I know my purpose in life which supports my "true" way.
Balancing Modality: Visualize the color orange. Say 2 times aloud, "I am impartial." Then, say 3 times aloud, "I know my purpose in life which supports my 'true' way."

2.4: The Soul

Color: Blue
Meridian: Liver, point #7, located on the inside of the left leg, just below the knee. (Chart 1)
Affirmation: I explore and am receptive to new ideas, methods, and opportunities.
Balancing Modality: Cover the left point #7 liver meridian with the right hand and say 2 times aloud, "I explore and am receptive to new ideas, methods, and opportunities." Visualize the color blue expanding from the left point #7 liver meridian throughout the entire body. Think of a time in the past when you were not receptive to new ideas, methods, and opportunities. Visualize a positive outcome to that experience by opening yourself to new possibilities. Surround the new outcome in the color blue.

2.5: Youth

Color: Red
Meridian: Large intestines, point #15, located on the left shoulder where the arm meets the shoulder. (Chart 3)
Affirmation: I am perfect.
Balancing Modality: Tone the note of F# 9 times, sending the note out of your body through the point #15 large intestines meridian. You will probably need to use a pitch pipe or musical instrument to assist you in locating the note. Say 4 times aloud, "I am perfect." Feel that sense of perfection enter your body through the left point #15 and spread throughout your body.

2.6: Independence/Interdependence

Color: Green
Meridian: Bladder, point #45, in middle of back, just above waist line on right side of spine. (Chart 2)
Affirmation: I have deep, reflective thoughts.
Balancing Modality: Hold your right hand over the right point #45 bladder meridian. Say 4 times aloud, "I have deep, reflective thoughts."

2.7: Growth

Color: Indigo
Meridian: Large intestines, point #15, located on the left shoulder where the arm meets the shoulder. (Chart 3)
Affirmation: I share my values without imposing them on others.
Balancing Modality: Breathe in deeply through your nose, filling your body. Breathe out through your mouth, making a long "Haaaaa" sighing sound. Repeat this breath twice again.

2.M: Maturity

Color: White

Meridian: Spleen, point #20, located at the end of the collar bone where the right arm adjoins to the shoulder. (Chart 1)

Affirmation: I clearly see where I am going.

Balancing Modality: Place your right hand over the left point #20 spleen meridian. Breath in deeply through your nose, filling your body. Breathe out through your mouth, making a long "Haaaaa" sighing sound. As you release this breath, push the color white through the left point #20 spleen meridian. Repeat this procedure once again.

Strand 3 - Creation

3.1: Pre-birth
Color: Magenta
Meridian: Heart, point # 3, located on the inside of the left arm, just above the elbow. (Chart 1)
Affirmation: I encompass all discordant elements when creating so that each can contribute to the end result according to its nature.
Balancing Modality: Place your right thumb on the side of your right nostril to block the air flow. Exhale and inhale through your left nostril. Then, place your right forefinger on the left nostril. Exhale and inhale through your right nostril. Alternate nostrils 4 times, completing 2 cycles of breathing.

3.2: Soul Entry
Color: Violet
Meridian: Circulation-sex, point #2, located on the right biceps, slightly higher than the nipple. (Chart 1)
Affirmation: I am in touch with my own feelings.
Balancing Modality: Say aloud 3 times, "I am in touch with my own feelings." Complete 1 breath as follows: Let the lower abdomen move inward as you inhale. Hold your breath, soften your eyes, and bring your head back, then turn your head to the right, then turn your head to the left, and finally breathe out half of the air you are holding through your nose while facing left. Now exhale the remainder of the air you are holding through your mouth, blowing out in a gust of wind. Visualize the color violet expanding in the #2 pelvic chakra (below the navel and above the pubic bone), your creation/sexual chakra. Hold the visualization until you feel a surge of creative energy emitting from the area.

3.3: Instinct
Color: Orange
Meridian: Spleen, point #8, on the inside of the left calf about four inches below the knee. (Chart 1)
Affirmation: I am self-assured.
Balancing Modality: Say aloud 4 times, "I am self-assured." Visualize the color orange going into the left point #8 spleen meridian. Place your right thumb on the side of your right nostril to block the air flow. Exhale and inhale through your left nostril. Then, place your right forefinger on the left nostril. Exhale and inhale through your right nostril. Alternate nostrils 8 times, completing 4 cycles of breathing.

3.4: The Soul

Color: Blue
Meridian: Lung, point #3, located on the right biceps even with the nipple line. (Chart 1)
Affirmation: My mind is precise.
Balancing Modality: Say aloud 2 times, "My mind is precise." Visualize the color blue going into the right point #3 lung meridian and traveling through your body to the #2 pelvic chakra (below the navel and above the pubic bone), your creation/sexual chakra. Stir the energy of that area by stimulating it with the color blue.

3.5: Youth

Color: Red
Meridian: Kidney, point #24, on the left side of the sternum (upper chest), approximately a hand's width down and directly underneath the beginning of your collar bone. (Chart 1)
Affirmation: I regulate and direct the flow of my creativity.
Balancing Modality: Hold your right hand over your right mu point for the liver (located under the ribs in line with the right nipple). Say aloud two times, "I regulate and direct the flow of my creativity."

3.6: Independence/Interdependence

Color: Green
Meridian: Large intestines, point #17, located just above the left collarbone at the base of the neck. (Chart 3)
Affirmation: I release worn out beliefs.
Balancing Modality: Do the following meridian massage: Lightly stroke up the outside of your right arm with your left hand 3 times, starting at the tips of your fingers and brushing up over your shoulders and up your neck and the side of your face. Repeat this using your right hand and brushing up your left arm. Visualize your old beliefs in the form of a green mass. Pack those old beliefs with their limitations into a hard ball and pitch that ball into the heavens and watch it disappear. Say aloud 4 times, "I release worn out beliefs."

3.7: Growth

Color: Indigo
Meridian: Bladder, point #3, located above the left eye at the top of the forehead where the forehead meets the head. (Chart 3)
Affirmation: I have energy reserves that I share and contain appropriately.
Balancing Modality: Breathe in deeply through your nose, filling your body. Breathe out through your mouth, making a long "Haaaaa" sighing sound.

Repeat this breath twice again. Place your left hand over the left point #3 bladder meridian. Say 2 times aloud, "I have energy reserves that I share and contain appropriately." Again, breathe in deeply through your nose, filling your body. Breathe out through your mouth, making a long "Haaaaa" sighing sound. Repeat this breath 2 more times.

3.M: Maturity

Color: White

Meridian: Liver, point #9, located on the inside of the left leg, slightly higher than midway between the knee and the crotch. (Chart 1)

Affirmation: I welcome transformation in myself and others.

Balancing Modality: Say aloud 9 times, "I welcome transformation in myself and others." Place your right hand over your heart. Then place your left hand over your right hand. Blink slowly 10 times.

Strand 4 - Well-being

4.1: Pre-birth
Color: Magenta
Meridian: Heart, point #5, located on the pinkie side of the left hand, 3 fingers above the wrist bone. (Chart 1)
Affirmation: I am open to everybody and every situation.
Balancing Modality: Place the right hand over the left point #5 heart meridian. Say 4 times aloud, "I am open to everybody and every situation." Hold your right hand over your right mu point for the liver (located under the ribs in line with the right nipple). Visualize the color magenta entering that mu point. As the color enters, feel yourself be receptive to every person and situation. Hold that feeling as you tone the "ore" sound, hollowing your mouth as if you have a small golf ball at the back of your throat to create resonance and touching your tongue to the roof of your mouth. Inhale deeply through your nose and exhale with your tongue touching the roof of your mouth, making a "SSSSS" sound like a snake. Repeat 4 times.

4.2: Soul Entry
Color: Violet
Meridian: Triple heater, point #2, located at the back (outside) of the left ring finger just above where the finger meets the palm. (Chart 2)
Affirmation: I bring harmony and balance to relationships.
Balancing Modality: Say 9 times aloud, "I bring harmony and balance to relationships." Complete 3 breaths as follows: Let the lower abdomen move inward as you inhale. Hold your breath, soften your eyes, and bring your head back, then turn your head to the right, then turn your head to the left, and finally breathe out half of the air you are holding through your nose while facing left. Now exhale the remainder of the air you are holding through your mouth, blowing out in a gust of wind. Look straight ahead as you repeat this breath two more times.

4.3: Instinct
Color: Orange
Meridian: Heart, point #7, located on pinkie side of left hand at point where the hand meets the wrist. (Chart 1)
Affirmation: I have clear insight which allows me to integrate all differences.
Balancing Modality: Inhale through your nose. Exhale through your mouth. Now wait for the body to determine when it needs to inhale. While waiting, relax and repeat the affirmation "I have clear insight which allows me to

integrate all differences." Visualize the color orange entering the left heart point #7. As it enters, send it through your body to the root chakra (located at the base of your spine). Let the color accumulate in the root chakra area until you feel as if you are sitting in a pool of orange. Feel your integration with all things are you visualize a web of longitude and latitude lines protruding from your body, linking with the universe above and the Earth around you.

4.4: The Soul

Color: Blue
Meridian: Triple heater, point #19, located on the left side of the head, directly behind the upper part of the ear, in line with your left eye. (Chart 3)
Affirmation: I create the right atmosphere for any activity.
Balancing Modality: Complete 4 breaths as follows: Let the lower abdomen move inward as you inhale. Hold your breath, soften your eyes, and bring your head back, then turn your head to the right, then turn your head to the left, and finally breathe out half of the air you are holding through your nose while facing left. Exhale the remainder of the air you are holding through your mouth, blowing out in a gust of wind. Look straight ahead as you repeat this breath three times. Say once aloud, "I create the right atmosphere for any activity."

4.5: Youth

Color: Red
Meridian: Lung, point #5, located just above the crook of the left arm below the biceps. (Chart 1)
Affirmation: I inspire others.
Balancing Modality: Say 2 times aloud, "I inspire others." Place your right hand over the left point #5 lung meridian. Say 2 times aloud again, "I inspire others." Complete 2 breaths as follows: Let the lower abdomen move inward as you inhale. Hold your breath, soften your eyes, and bring your head back, then turn your head to the right, then turn your head to the left, and finally breathe out half of the air you are holding through your nose while facing left. Now exhale the remainder of the air you are holding through your mouth, blowing out in a gust of wind. Look straight ahead as you repeat this breath once again.

4.6: Independence/Interdependence

Color: Green
Meridian: Triple heater, point #23, located on the left temple, just beyond the eyebrow and slightly above the eye. (Chart 3)

Affirmation: I am loving.
Balancing Modality: Say 2 times aloud, "I am loving." Place your left hand over the left point #23 triple heater meridian. Say 2 times aloud again, "I am loving."

4.7: Growth
Color: Indigo
Meridian: Heart, point #6, located on the pinkie side of the left hand, 2 fingers above the wrist bone. (Chart 1)
Affirmation: I treat everyone with love and understanding.
Balancing Modality: Complete 2 breaths as follows: Let the lower abdomen move inward as you inhale. Hold your breath, soften your eyes, and bring your head back, then turn your head to the right, then turn your head to the left, and finally breathe out half of the air you are holding through your nose while facing left. Now exhale the remainder of the air you are holding through your mouth, blowing out in a gust of wind. Look straight ahead as you repeat this breath again. Say 2 times aloud, "I treat everyone with love and understanding." Place your thumbs on your temples. Feel a sense of love coming into your temples from the Divine Creator. Extend that feeling to all others in your life.

4.M: Maturity
Color: White
Meridian: Gallbladder, point #25, located on the left side of the abdomen at waist level, in line with the armpit. (Chart 3)
Affirmation: I know who I am and where I stand.
Balancing Modality: Visualize the color white entering the left gallbladder point #25 while saying 1 time aloud "I know who I am and where I stand." Feel the certainty that comes with that knowing.

Strand 5 - Joy

5.1: Pre-birth

Color: Magenta

Meridian: Small intestines, point #8, located directly underneath the right elbow. (Chart 2)

Affirmation: I am courteous.

Balancing Modality: Do the following meridian massage: Lightly stroke down the inside of your right arm with your left hand 3 times from the armpit to the fingertips. Repeat this using your right hand and brushing down your left arm. Lightly stoke up the outside of the right arm with your left hand 3 times, starting at the tips of your fingers and brushing up over your shoulders and up your neck and the side of your face. Repeat this using your right hand and brushing up your left arm. Now, place your hands over your eyes with fingers spread wide and brush your hands over your head, down the back of your neck as far as you can reach, then pick it up where you left off and stroke down your back and buttocks. Continue stroking down the outside of your legs, rounding the outside of your feet to your little then big toes. Finish by stroking up the inside of your feet and legs, over the genitals and up the center of your body to your armpits. Say once aloud, "I am courteous." Breathe in deeply through your nose, filling your body. Breathe out through your mouth, making a long "Haaaaa" sighing sound. Repeat this breath 3 more times.

5.2: Soul Entry

Color: Violet

Meridian: Triple heater, point #12, located midway between the armpit and the elbow on the back of the right arm. (Chart 2)

Affirmation: I maintain the right temperature for the optimal flow of any activity.

Balancing Modality: Say 2 times aloud, "I maintain the right temperature for the optimal flow of any activity." Inhale deeply through your nose and exhale with your tongue against the roof of your mouth, making a long "SSSSS" sound like a snake. Repeat this 5 times. Place your left hand over the right point #12 triple heater meridian. Begin a continuous breath, inhaling and exhaling without pausing at the bottom of the exhale or the top of the inhale. Repeat this continuous breath for 10 cycles.

5.3: Instinct
Color: Orange
Meridian: Circulation-sex, point #4, located on the left inside forearm, midway between the crook of the arm and the wrist, somewhat closer to the wrist. (Chart 1)
Affirmation: I appropriately confide in others.
Balancing Modality: Place your right hand over the left point #4 circulation-sex meridian, and say 2 times aloud, "I appropriately confide in others." Visualize the color orange expanding in the left point #4 circulation-sex meridian area. Allow the orange color to move into the #2 pelvic chakra (below the navel and above the pubic bone), your creation/sexual chakra. Complete 1 breath as follows: exhale with a "Ha" sound, then inhale. Exhale with a "Whoo" sound, then inhale. Exhale with a "Hee" sound, then inhale. Exhale with a "Tsu" sound, then inhale. Exhale with a "Ssss" sound, then inhale.

5.4: The Soul
Color: Blue
Meridian: Spleen, point #20, located at the end of the collar bone where the right arm adjoins to the shoulder. (Chart 1)
Affirmation: I live in a state of peace and joy.
Balancing Modality: Say 1 time aloud, "I live in a state of peace and joy." Allow yourself to remember an earlier time, prior to coming to Earth, when you lived in a constant state of peace and joy. Feel that ever-present feeling of freedom that is derived from continual peace and joy. Allow that feeling to merge with the color blue and disperse it to the atmosphere through your right point #20 spleen meridian.

5.5: Youth
Color: Red
Meridian: Lung, point #10, located on the left hand where the thumb meets the back of the hand. (Chart 1)
Affirmation: I inspire others through example.
Balancing Modality: Breathe in deeply through your nose, filling your body. Breathe out through your mouth, making a long "Haaaaa" sighing sound. Repeat this breath 5 times. Say 1 time aloud, "I inspire others through example."

5.6: Independence/Interdependence

Color: Green
Meridian: Bladder, point #54, located at the back of the right knee in the bend of the leg. (Chart 2)
Affirmation: I have deep reserves of energy that I share and contain appropriately.
Balancing Modality: Inhale deeply through your nose and exhale with your tongue against the roof of your mouth, making a long "SSSSS" sound like a snake. Repeat this breath 7 times. Place your right hand over the right point #54 kidney meridian. Say 4 times out loud, "I have deep reserves of energy that I share and contain appropriately." Visualize the color green holding those deep reserves of energy. Place that green bubble of energy in your heart chakra.

5.7: Growth

Color: Indigo
Meridian: Spleen, point #21, located on the left side of the chest near the armpit on level with the sixth rib from the top. (Chart 1)
Affirmation: I am at peace.
Balancing Modality: Place your right hand over the left point #21 spleen meridian. Say 19 times aloud, "I am at peace." Feel the peace as you state your intention. Visualize that peaceful feeling in the form of the color indigo, and let it disperse throughout your body.

5.M: Maturity

Color: White
Meridian: Small intestines, point #19, located on the right side of the face at the top of and behind the cheekbone in front of the ear. (Chart 1)
Affirmation: I extract what is pure in any situation.
Balancing Modality: Inhale deeply through your nose and exhale with your tongue against the roof of your mouth, making a long "SSSSS" sound like a snake. Repeat this breath 8 times. Say 2 times aloud, "I extract what is pure in any situation." Visualize the color white enveloping your heart as you experience this purity.

Strand 6 - Truth

6.1: Pre-birth
Color: Magenta
Meridian: Small intestines, point #12, touch left shoulder midway between neck and arm and slide hand down over the top portion of the back. You will be covering point #12. (Chart 2)
Affirmation: My humor is warm and loving.
Balancing Modality: Say 3 times aloud, "My humor is warm and loving." Visualize the color magenta surrounding your body. Bring the color into your body through the heart chakra.

6.2: Soul Entry
Color: Violet
Meridian: Small intestines, point #12, touch left shoulder midway between neck and arm and slide hand down over the top portion of the back. You will be covering point #12. (Chart 2)
Affirmation: I enjoy sexual intimacy based on truth for the bonding it brings.
Balancing Modality: Complete 3 breaths as follows: Let the lower abdomen move inward as you inhale. Hold your breath, soften your eyes, and bring your head back, then turn your head to the right, then turn your head to the left, and finally breathe out half of the air you are holding through your nose while facing left. Now exhale the remainder of the air you are holding through your mouth, blowing out in a gust of wind. Look straight ahead as you repeat this breath 2 more times. Then, say 2 times aloud, "I enjoy sexual intimacy based on truth for the bonding it brings."

6.3: Instinct
Color: Orange
Meridian: Small intestines, point #12, touch left shoulder midway between neck and arm and slide hand down over the top portion of the back. You will be covering point #12. (Chart 2)
Affirmation: I sort out what is best and resolve all splits between approaches.
Balancing Modality: Place your right thumb on the side of your right nostril to block the air flow. Exhale and inhale through your left nostril. Then, place your right forefinger on the left nostril. Exhale and inhale through your right nostril. Alternate nostrils 8 times, completing 4 cycles of breathing. Visualize the color orange entering the left point #12 small intestines meridian. As

it enters, say 2 times aloud, "I sort out what is best and resolve all splits between approaches."

6.4: The Soul

Color: Blue
Meridian: Heart, point #3, located on the inside of the left arm, just above the elbow. (Chart 1)
Affirmation: I have a strong heart.
Balancing Modality: Say 4 times aloud, "I have a strong heart." Release the fear of being yourself by formulating the feeling, then dispersing it as you continually move your eyes diagonally from the upper left to the lower right until all of that feeling is released. Place your right hand over the left point #3 heart meridian. Visualize the color blue expanding in this area.

6.5: Youth

Color: Red
Meridian: Circulation-sex, point #5, located on the left inside forearm, approximately a hand's width from the wrist. (Chart 1)
Affirmation: I have intimate heart-to-heart connections.
Balancing Modality: Say 3 times aloud, "I have intimate heart-to-heart connections." Visualize the color red at the alpha chakra (located in the middle thigh area).

6.6: Independence/Interdependence

Color: Green
Meridian: Bladder, point #22, located on the right side of the spinal cord, just above the waist. (Chart 2)
Affirmation: I am reflective.
Balancing Modality: Place your right thumb on the side of your right nostril to block the air flow. Exhale and inhale through your left nostril. Then, place your right forefinger on the left nostril. Exhale and inhale through your right nostril. Alternate nostrils 4 times, completing 2 cycles of breathing. Say 2 times aloud, "I am reflective." Visualize the color green expanding in the throat chakra.

6.7: Growth
Color: Indigo

Meridian: Lung, point #7, located on the right arm in line with the thumb, just above the wrist. (Chart 1)
Affirmation: I have pride in my family.

Balancing Modality: Bend over and position the palms of your hands a foot away but facing your body. Imagine a blue light as you sweep your hands slowly up your body, stopping at the heart chakra. Let the blue light fill your heart. Sweep your hands over your head and down the back of your body and legs. Now, imagine an indigo light as you sweep your hands slowly up your body, stopping at the brow chakra (located in the middle of your forehead at your third eye). Let the indigo light fill your brow chakra. Sweep your hands over your head and down the back of your body and legs.

6.M: Maturity

Color: White

Meridian: Spleen, point #20, located at the end of the collar bone where the right arm adjoins to the shoulder. (Chart 1)

Affirmation: I am vital.

Balancing Modality: Say 2 times aloud, "I am vital." Place your left hand over your right point #20 spleen meridian. Tone the note of C# 10 times, sending the note out of your body through the right point #20 spleen meridian. You will probably need to use a pitch pipe or musical instrument to assist you in locating the note. Breathe in deeply through your nose, filling your body. Breathe out through your mouth, making a long "Haaaaa" sighing sound. Repeat this breath 6 times.

Strand 7 - Balance

7.1: Pre-birth
Color: Magenta
Meridian: Heart, point #7, located on pinkie side of left hand at point where the hand meets the wrist. (Chart 1)
Affirmation: I eliminate the resistance between my heart and the hearts of others.
Balancing Modality: Inhale deeply through your nose and exhale with your tongue against the roof of your mouth, making a long "SSSSS" sound like a snake. Repeat this breath 5 times. Say 2 times aloud, "I eliminate the resistance between my heart and the hearts of others."

7.2: Soul Entry
Color: Violet
Meridian: Stomach, point #45, at the end of the right middle toe. (Chart 1)
Affirmation: I am content and at peace.
Balancing Modality: Say 4 times aloud, "I am content and at peace." Visualize a violet pool of water in which you are floating effortlessly on the top. The violet represents contentment and peace. Feel the contentment and peace seep through you as you float on top of it.

7.3: Instinct
Color: Orange
Meridian: Triple heater, point #16, located on the back of the neck behind the left SCM muscle (on side of neck), midway between the hairline and the shoulders. (Chart 3)
Affirmation: All relationships are important to me.
Balancing Modality: Say 4 times aloud, "All relationships are important to me." Inhale fully and, without pausing, exhale. Then exhale fully and, without pausing, inhale. Continue inhaling and exhaling without pausing for several minutes.

7.4: The Soul
Color: Blue
Meridian: Spleen, point #21, located on the left side of the chest near the armpit on level with the sixth rib from the top. (Chart 1)
Affirmation: I act responsibly.
Balancing Modality: Say aloud 7 times, "I act responsibly." Place your right hand over the left point #21 spleen meridian. Tap your thymus (located in

the middle of your chest, midway between the collar bone and the breast) 30 times.

7.5 Youth
Color: Red
Meridian: Circulation-sex, point #8, located in the middle of the left palm. (Chart 1)
Affirmation: I circulate love.
Balancing Modality: Inhale through your nose. Exhale through your mouth. Now wait for the body to determine when it needs to repeat the inhale. While waiting, relax and let go, repeating the affirmation "I circulate love."

7.6: Independence/Interdependence
Color: Green
Meridian: Large intestines, point #19, above upper right lip. (Chart 1)
Affirmation: I am aware of my self-righteousness, and I do something about it.
Balancing Modality: Say 10 times aloud, "I am aware of my self-righteousness, and I do something about it." Take in the feeling of humility, knowing that you are Divine but that you are also humble in relationship to the Divine Creator who is all-knowing. Complete 1 breath as follows: exhale with a "Haa" sound, then inhale. Exhale with a "Whoo" sound, then inhale. Exhale with a "Hee" sound, then inhale. Exhale with a "Tsu" sound, then inhale. Exhale with a "Ssss" sound, then inhale.

7.7: Growth
Color: Indigo
Meridian: Liver, point #14, located about 4 inches under right nipple in line with right nipple. (Chart 1)
Affirmation: I envision my spiritual ideal and do those actions that support my ideal.
Balancing Modality: Remember when you were a child and had ideas about what you wanted to be when you grew up. You were sure you could become a prima ballerina or a concert pianist or a famous baseball player. It never occurred to you that you could not be something you wanted. Hold that feeling of realization of your goals as you place your left hand over the right point #14 liver meridian. Say twice aloud, "I envision my spiritual ideal and do those actions that support my ideal."

7.M: Maturity

Color: White

Meridian: Heart, point #8, located on pinkie side of left hand, midway between the wrist and the base of the pinkie. (Chart 1)

Affirmation: I project a positive influence.

Balancing Modality: Inhale deeply through your nose and hold your breath. Push air gently into your lower abdomen. While holding the air in your abdomen, squeeze and release your sphincter muscle 2 times. Then exhale through your nose.

Strand 8 - Power

8.1: Pre-birth
Color: Magenta
Meridian: Liver, point #13, located under rib under right nipple in line with right nipple midway between the waist line and the nipple. (Chart 1)
Affirmation: I create the strategies for manifesting my plans and goals.
Balancing Modality: Say 3 times aloud, "I create the strategies for manifesting my plans and goals." Breathe in through your nose, drawing in a feeling of creativity. Exhale and let go of any blocks to creativity.

8.2: Soul Entry
Color: Violet
Meridian: Heart, point #7, located on pinkie side of left hand at point where the hand meets the wrist. (Chart 1)
Affirmation: I know all aspects of myself, both light and dark, and fully appreciate who I am.
Balancing Modality: Say 7 times aloud, "I know all aspects of myself, both light and dark, and fully appreciate who I am." Place your right hand over the left point #7 heart meridian. Then cover your heart chakra with your left hand while your right hand is still in place. Look up, right, down, and left with your eyes. Repeat this eye movement sequence 2 times.

8.3: Instinct
Color: Orange
Meridian: Liver, point #13, located under rib under right nipple in line with right nipple midway between the waist line and the nipple. (Chart 1)
Affirmation: I envision my spiritual ideal and do those actions that support that ideal.
Balancing Modality: Say 3 times aloud, "I envision my spiritual ideal and do those actions that support that ideal." Make a cross with your eyes by looking up, down, left, and right. Focus ahead and visualize the color orange glowing in the distance, as the sun would rise on the horizon. See your spiritual ideals coming to fruition as the orange explodes over the horizon.

8.4: The Soul
Color: Blue
Meridian: Triple heater, point #16, located on the back of the neck behind the left SCM muscle (on side of neck), midway between the hairline and the shoulders. (Chart 3)
Affirmation: I keep things in proportion and see the big picture.
Balancing Modality: Exhale strongly through your nose, squeezing your stomach in. Inhale deeply through your nose. Repeat this sequence 4 times. Visualize the color blue entering the left point #16 triple heater meridian.

8.5: Youth
Color: Red
Meridian: Small intestines, point #19, located on the right side of the face at the top of and behind the cheekbone in front of the ear. (Chart 1)
Affirmation: I allow for the development of the finer qualities of myself.
Balancing Modality: Visualize a rainbow extending over you. This rainbow is composed of all of your finer qualities. Watch as people cluster around you to admire the rainbow, knowing they are viewing the best that you have to offer. Say once aloud, "I allow for the development of the finer qualities of myself." Watch as the rainbow grows and expands until it encircles your neighborhood, city, state, continent, world, and, finally, the universe.

8.6: Independence/Interdependence
Color: Green
Meridian: Gall bladder, point #20, located at the left back of the neck where the hairline meets the spine. (Chart 2)
Affirmation: I come from integrity.
Balancing Modality: Inhale through your nose. Exhale through your mouth. Now wait for the body to determine when it needs to repeat the inhale. Repeat this sequence 4 more times. Say 2 times aloud, "I come from integrity."

8.7: Growth
Color: Indigo
Meridian: Small intestines, point #19, located on the right side of the face at the top of and behind the cheekbone in front of the ear. (Chart 1)
Affirmation: I enjoy sexual intimacy for the bonding and joy it brings.
Balancing Modality: Place your left hand over the right point #19 small intestines meridian. Visualize making love with the perfect life mate, one who brings great joy to you through the heart connection that binds you. Say 3 times aloud, "I enjoy sexual intimacy for the bonding and joy it brings."

8.M: Maturity

Color: White

Meridian: Lung, point #11, located at the tip of the left thumb. (Chart 1)

Affirmation: I take pride in my planetary family for each is acting according to their own Divine plan.

Balancing Modality: Focus on something near to you. Then focus on something far away. Repeat this sequence 2 more times. Say 2 times aloud, "I take pride in my planetary family for each is acting according to their own Divine plan." Make a cross with your eyes by looking up, down, left, and right.

Strand 9 - Communion

9.1: Pre-birth

Color: Magenta
Meridian: Small intestines, point #16, located behind the right SCM muscle (on side of neck) on level with the Adam's apple. (Chart 1)
Affirmation: I resolve all splits between me and my partner.
Balancing Modality: Say 8 times aloud, "I resolve all splits between me and my partner." Visualize the color magenta entering the right point #16 small intestines meridian. Allow the color to fill your entire body. Tone the note of D 5 times. You will probably need to use a pitch pipe or musical instrument to assist you in locating the note.

9.2: Soul Entry

Color: Violet
Meridian: Small intestines, point #16, located behind the right SCM muscle (on side of neck) on level with the Adam's apple. (Chart 1)
Affirmation: My humor is warm and loving.
Balancing Modality: Breathe in deeply through your nose, filling your body. Breathe out through your mouth, making a long "Haaaaa" sighing sound. Repeat this breath 4 times. Say 2 times aloud, "My humor is warm and loving." Place your right hand over the right point #16 small intestines meridian. Visualize the joyful response of others when you use your humor lovingly rather than at the expense of others.

9.3: Instinct

Color: Orange
Meridian: Triple heater, point #19, located on the left side of the head, directly behind the upper part of the ear, in line with your left eye. (Chart 3)
Affirmation: All relationships are important to me.
Balancing Modality: Place your right hand over the left point #19 triple heater meridian. Tone the note of C 2 times. You will probably need to use a pitch pipe or musical instrument to assist you in locating the note.

9.4: The Soul
Color: Blue
Meridian: Circulation-sex, point #7, located on the inside of the right arm where the hand meets the wrist. (Chart 1)
Affirmation: I circulate love.
Balancing Modality: Say 2 times aloud, "I circulate love." Visualize the color blue surrounding your entire body as you circulate your love, sending the blue out in waves to everyone and everything in the universe.

9.5: Youth
Color: Red
Meridian: Spleen, point #20, located at the end of the collar bone where the right arm adjoins to the shoulder. (Chart 1)
Affirmation: I know where I am going.
Balancing Modality: Say 2 times aloud, "I know where I am going." Inhale deeply through your nose and exhale with your tongue against the roof of your mouth, making a long "SSSSS" sound like a snake. Repeat this breath 3 times. Visualize the color red encircling your head like a ring around your forehead. Feel the energy circulating clockwise around the ring, feeling a sense of certainty being grounded regarding your confidence in where you are going.

9.6: Independence/Interdependence
Color: Green
Meridian: Circulation-sex, point #8, located in the middle of the left palm.
Affirmation: I have the capacity for a lifelong love commitment. (Chart 1)
Balancing Modality: Place your right hand over the left point #8 circulation-sex meridian. Say 2 times aloud, "I have the capacity for a lifelong love commitment. Place your right thumb on the side of your right nostril to block the air flow. Exhale and inhale through your left nostril. Then, place your right forefinger on the left nostril. Exhale and inhale through your right nostril. Alternate nostrils 6 times, completing 3 cycles of breathing.

9.7: Growth
Color: Indigo
Meridian: Spleen, point #21, located on the left side of the chest near the armpit on level with the sixth rib from the top. (Chart 1)
Affirmation: I am at peace.
Balancing Modality: Place your left hand over your left point #21 spleen meridian. Inhale fully and, without pausing, exhale. Then exhale fully and, without pausing, inhale. Continue inhaling and exhaling without pausing for several minutes. Then say 3 times aloud, "I am at peace."

9.M: Maturity

Color: White

Meridian: Kidney, point #26, located under the neck, just below the left collarbone. (Chart 1)

Affirmation: I easily accept changes in direction.

Balancing Modality: Say 5 times aloud, "I easily accept changes in direction." Visualize the color white descending from above and entering your crown chakra. Imagine all of the changes that will be required of you in the future, changes that will be easy to accommodate by your acceptance of them. Feel at peace with yourself, knowing you have the ability to move with these changes by not resisting them.

Strand 10 - Cohesiveness

10.1: Pre-birth

Color: Magenta

Meridian: Lung, point #7, located on the right arm in line with the thumb, just above the wrist. (Chart 1)

Affirmation: I take pride in my family.

Balancing Modality: Say 4 times aloud, "I take pride in my family." Complete 1 breath as follows: exhale with a "Ha" sound, then inhale. Exhale with a "Whoo" sound, then inhale. Exhale with a "Hee" sound, then inhale. Exhale with a "Tsu" sound, then inhale. Exhale with a "Ssss" sound, then inhale. Repeat this breath. Visualize that you are standing in a pool of the color magenta. Feel its cooling refreshment as you splash through the colored pool.

10.2: Soul Entry

Color: Violet

Meridian: Gall bladder, point #24, located approximately 5" below left nipple in line with the nipple. (Chart 3)

Affirmation: I move forward with vision and make the necessary changes that are needed.

Balancing Modality: Cover the left point #24 gall bladder meridian with the right hand. Visualize the color violet entering this meridian point and descending to the feet, exiting the soles of your feet and entering the Earth. Now sense yourself moving forward with vision and making the necessary changes that are needed while you are grounded by the violet threads of light. Exhale all the air from your lungs and relax deeply. Without breathing, imagine that you are breathing internally, obtaining your oxygen from within. Envision a life made easy by your readiness to make the necessary changes along your path.

10.3: Instinct

Color: Orange

Meridian: Liver, point #14, located about 4 inches under right nipple in line with right nipple. (Chart 1)

Affirmation: My vision and creativity allow my ideas to flow.

Balancing Modality: Say 4 times aloud, "My vision and creativity allow my ideas to flow." Visualize the color orange as a cloud of color that holds all of your creative ideas. See the cloud grow and expand as you add more and more ideas.

10.4: The Soul
Color: Blue
Meridian: Small intestines, point #14, touch left shoulder midway between neck and arm and slide hand down over the top portion of the back. You will be covering point #14. (Chart 2)
Affirmation: My language is clean and strengthening.
Balancing Modality: Say 3 times aloud, "My language is clean and strengthening." Surround yourself in red light and breath in and out several times. Then surround yourself in orange light and breath in and out several times. Repeat this with yellow, green, blue, indigo, violet, and white light. Complete 1 breath as follows: exhale with a "Ha" sound, then inhale. Exhale with a "Whoo" sound, then inhale. Exhale with a "Hee" sound, then inhale. Exhale with a "Tsu" sound, then inhale. Exhale with a "Ssss" sound, then inhale.

10.5: Youth
Color: Red
Meridian: Small intestines, point #19, located on the right side of the face at the top of and behind the cheekbone in front of the ear. (Chart 1)
Affirmation: I take the riches out of life for the full blossoming of my being.
Balancing Modality: Say 3 times aloud, "I take the riches out of life for the full blossoming of my being." Visualize the color red surrounding your entire body, as if you are sitting inside a red egg. Complete 1 breath as follows: exhale with a "Ha" sound, then inhale. Exhale with a "Whoo" sound, then inhale. Exhale with a "Hee" sound, then inhale. Exhale with a "Tsu" sound, then inhale. Exhale with a "Ssss" sound, then inhale. Complete a second round of sounds and breath.

10.6: Independence/Interdependence
Color: Green
Meridian: Stomach, point #45, at the end of the right middle toe. (Chart 1)
Affirmation: I am clear headed.
Balancing Modality: Inhale deeply through your nose and hold your breath. Push air gently into your lower abdomen. While holding the air in your abdomen, squeeze and release your sphincter muscle 3 times. Then exhale through your nose. Repeat this breath 2 times. Cover your right point #45 stomach meridian with your right hand. Say 3 times aloud, "I am clear headed."

10.7: Growth

Color: Indigo
Meridian: Heart, point #7, located on pinkie side of left hand at point where the hand meets the wrist. (Chart 1)
Affirmation: I understand each person's needs.
Balancing Modality: Say 3 times aloud, "I understand each person's needs." Think back to a time in your life when you put your own needs in front of another person's needs because you did not understand what they desired. Visualize a different outcome based on hindsight which has given you greater understanding in terms of what the other needed at the time. Remind yourself that you always have the capacity to view a situation from another's perspective.

10.M: Maturity

Color: White
Meridian: Spleen, point #20, located at the end of the collar bone where the right arm adjoins to the shoulder. (Chart 1)
Affirmation: I am responsible.
Balancing Modality: Say 3 times aloud, "I am responsible." Cover the right point #20 spleen meridian with your left hand. Inhale deeply through your nose and hold your breath. Push air gently into your lower abdomen. While holding the air in your abdomen, squeeze and release your sphincter muscle 5 times. Then exhale through your nose. Repeat this breath 4 times. Visualize the color white throughout your entire body. Feel it purifying you.

Strand 11 - Sexuality

11.1: Pre-birth
Color: Magenta
Meridian: Heart, point #4, located on the pinkie side of the left hand, 4 finger widths above the wrist bone. (Chart 1)
Affirmation: I transcend all differences to give fully as appropriate.
Balancing Modality: Breath in deeply through your nose, filling your body. Breathe out through your mouth, making a long "Haaaaa" sighing sound. Repeat this breath once again. Hold your right hand over left point #4 heart meridian and say 4 times aloud, "I transcend all differences to give fully as appropriate."

11.2: Soul Entry
Color: Violet
Meridian: Small intestines, point #18, located directly on the right cheek-bone. (Chart 1)
Affirmation: I determine what is in the highest good of all concerned and resolve conflict between approaches.
Balancing Modality: Say 2 times aloud, "I determine what is in the highest good of all concerned and resolve conflict between approaches." Visualize the color violet entering the right point #18 small intestines meridian as you feel your creativity enhanced rather than limited by situations involving conflicting points of view.

11.3: Instinct
Color: Orange
Meridian: Lung, point #5, located on the right inner arm, just above the crook of the arm. (Chart 1)
Affirmation: I have pride in my family.
Balancing Modality: Hold your left hand over the right point #5 lung meridian. Inhale sharply through your nose, puffing out your chest and hold the breath. Exhale with force through your mouth, imagining that you are pushing the air out through your feet. Repeat this breath once more. Say once aloud, "I have pride in my family."

11.4: The Soul
Color: Blue
Meridian: Spleen, point #21, located on the left side of the chest near the armpit on level with the sixth rib from the top. (Chart 1)
Affirmation: I am at peace.
Balancing Modality: Say once aloud, "I am at peace." Tone the "aahh" sound 5 times, hollowing your mouth as if a golf ball was sitting on your tongue to create resonance and placing the tip of your tongue on the roof of your mouth. Visualize the color blue seeping throughout your entire body, coming in through the crown and oozing through the feet.

11.5: Youth
Color: Red
Meridian: Lung, point #3, located on the right biceps even with the nipple line. (Chart 1)
Affirmation: I allow myself to experience my grief.
Balancing Modality: Hold your left hand over the right point #3 lung meridian. Breathe the feeling of grief into your body through point #3. As you feel grief, say once aloud, "I allow myself to experience my grief." Complete 2 breaths as follows: Let the lower abdomen move inward as you inhale. Hold your breath, soften your eyes, and bring your head back, then turn your head to the right, then turn your head to the left, and finally breathe out half of the air you are holding through your nose while facing left. Now exhale the remainder of the air you are holding through your mouth, blowing out in a gust of wind. Look straight ahead as you repeat this sequence.

11.6: Independence/Interdependence
Color: Green
Meridian: Liver, point #13, located under rib under right nipple in line with right nipple midway between the waist line and the nipple. (Chart 1)
Affirmation: I am receptive to new ideas.
Balancing Modality: Place your right thumb on the side of your right nostril to block the air flow. Exhale and inhale through your left nostril. Then, place your right forefinger on the left nostril. Exhale and inhale through your right nostril. Alternate nostrils 4 times, completing 2 cycles of breathing.

11.7: Growth

Color: Indigo
Meridian: Heart, point #2, located on the inside of the left arm, midway between the crook of the arm and the armpit, slightly closer to the crook of the arm. (Chart 1)
Affirmation: I have clear insight which allows me to integrate all differences.
Balancing Modality: Say twice aloud, "I have clear insight which allows me to integrate all differences. This is all that is necessary to balance this box.

11.M: Maturity

Color: White
Meridian: Bladder, point #10, at the base of the hairline on the left side of the neck. (Chart 3)
Affirmation: I have ideas and creativity and put them into action.
Balancing Modality: Lay down. Do not use a pillow to support your head. Visualize the color white entering the left point #10 bladder meridian until your entire head is filled with white light. Say once aloud, "I have ideas and creativity, and I put them into action."

Strand 12 - Love

12.1: Pre-birth

Color: Magenta

Meridian: Small intestines, point #16, located behind the right SCM muscle (on side of neck) on level with the Adam's apple. (Chart 1)

Affirmation: I am intellectually alive which results in my emotional well-being.

Balancing Modality: Hold your left hand over the right point #16 small intestines meridian. Examine one of the action plans you have devised in the past to increase your intellectual abilities. Trace those intellectual abilities to a situation that resulted in your emotional well-being. Feel the satisfaction that occurred. Feel the color magenta entering into point #16 small intestines meridian as you say 4 times aloud "I am intellectually alive which results in my emotional well-being."

12.2: Soul Entry

Color: Violet

Meridian: Heart, point # 3, located on the inside of the left arm, just above the elbow. (Chart 1)

Affirmation: I eliminate the resistance from one heart to another.

Balancing Modality: Think of someone you love, either alive or dead, and feel the heart connection, represented by a violet cord that extends from your heart to theirs. Next, place your right hand over the left point #3 heart meridian and say 4 times aloud, "I eliminate the resistance from one heart to another." Now, extend that feeling of openness to everyone in your life. Feel the inter-relatedness of all beings who are here in your life at this time. Extend that feeling to your town, to your state, to your country, and to the entire planet.

12.3: Instinct

Color: Orange

Meridian: Stomach, point #25, located about 1½ inches above and to the right of the navel. (Chart 1)

Affirmation: I appreciate and love others.

Balancing Modality: Tone the note of D# 4 times, sending the note out of your body through the right point #25 stomach meridian. You will probably need to use a pitch pipe or musical instrument to assist you in locating the note.

12.4: The Soul

Color: Blue
Meridian: Large intestines, point #15, located on the left shoulder where the arm meets the shoulder. (Chart 3)
Affirmation: I have a clear, sharp mind.
Balancing Modality: Let the color blue wash through your entire body as if it is being cleansed by the color. Say once aloud, "I have a clear, sharp mind." Complete 10 breaths as follows: Let the lower abdomen move inward as you inhale. Hold your breath, soften your eyes, and bring your head back, then turn your head to the right, then turn your head to the left, and finally breathe out half of the air you are holding through your nose while facing left. Now exhale the remainder of the air you are holding through your mouth, blowing it out in a gust of wind. Look straight ahead as you repeat this breath 9 more times.

12.5: Youth

Color: Red
Meridian: Large intestines, point #15, located on the left shoulder where the arm meets the shoulder. (Chart 3)
Affirmation: I am aware of my self-righteousness and do something about it.
Balancing Modality: Feel that feeling of self-righteousness when you are experiencing a situation when you are convinced you are right and someone else is wrong. See your self-righteousness as the color red. Watch the red dissipate as you recognize there is no right or wrong, only different experiences which create differing points of view. Feel the weight dissipate from your shoulders as you release the feeling of needing to be right. Say once aloud, "I am aware of my self-righteousness and do something about it."

12.6: Independence/Interdependence

Color: Green
Meridian: Triple heater, point #8, located on the back of the right forearm approximately one-fourth of the way above where the back of the hand meets the arm. (Chart 2)
Affirmation: All relationships are important to me.
Balancing Modality: Place your right thumb on the side of your right nostril to block the air flow. Exhale and inhale through your left nostril. Then, place your right forefinger on the left nostril. Exhale and inhale through your right nostril. Alternate nostrils 4 times, completing 2 cycles of breathing. Say 10 times aloud, "All relationships are important to me." Exhale all of the air out of your lungs and visualize the color green filling your lungs as you relax

deeply. Without breathing, imagine that you are breathing internally, obtaining your oxygen from the color green. Repeat this breath once again.

12.7: Growth

Color: Indigo

Meridian: Small intestines, point #12, touch left shoulder midway between neck and arm and slide hand down over the top portion of the back. You will be covering point #12. (Chart 2)

Affirmation: I am able to absorb what is valuable out of any situation.

Balancing Modality: Say aloud 10 times, "I am able to absorb what is valuable out of any situation." Place your right hand over the left point #12 small intestines meridian. Exhale all of the air out of your lungs and visualize the color indigo filling your lungs as you relax deeply. Without breathing, imagine that you are breathing internally, obtaining your oxygen from the color indigo.

12.M: Maturity

Color: White

Meridian: Triple heater, point #16, located on the left side of the neck approximately one inch behind and below the ear. (Chart 3)

Affirmation: I manifest an atmosphere around me so others feel like they fit.

Balancing Modality: Place your right thumb on the side of your right nostril to block the air flow. Exhale and inhale through your left nostril. Then, place your right forefinger on the left nostril. Exhale and inhale through your right nostril. Alternate nostrils 10 times, completing 5 cycles of breathing. Visualize the color white filling your entire body, oozing from your body so you are surrounded in a cloud of white light. Say aloud 3 times, "I manifest an atmosphere around me so others feel like they fit."

Appendix C

- Phase III -

Balancing Polarity

Polarity versus Divine Union

Almost one year after I finished phase one and two of the recoding process, I encountered a juncture that felt significantly different from the preceding peaceful year when I had established my union with David and evolved into a spiritual healer with a growing metaphysical practice. Prior to hitting this juncture, I had settled into a mind-set that was very comfortable. I knew who I was, I recognized my talents, I understood my soul path and purpose, and I awoke each day with an innate knowing of what I needed to accomplish. Suddenly, those feelings flew by the wayside. I lived for two months with an unsettled feeling of discomfort. Although I had not changed my lifestyle, I felt "off path and purpose" and stopped seeing clients, feeling as if I had nothing to offer. I seemed to lose a part of myself, repeating constantly to David that I did not know who I was anymore. This was very disturbing since, after previous years of questioning, I had grown very sure once I completed recoding. I could not reconcile my new feeling of disjointment with the supreme knowing I had enjoyed since recoding.

Additionally, I kept attracting negative energy into my life. Having mastered the basics of creative manifestation, I could not understand how I was responsible for the negative situations. I knew I had a lesson to learn, but I could not determine what it was. All I knew was the more loving I became, the more I seemed to attract unloving people into my life. The more generous I became, the more money difficulties I seemed to encounter. Finally, through a conversation with my mentor Venessa and subsequent channelings with my guides, I determined that I was being hampered in my path toward unification by soul contracts that were based on the premise of polarity. The more I achieved unification, the step beyond balance when resistance is nil, the more the mass consciousness of polarity attempted to impose itself by bringing the opposite situation.

The following information was provided by Joysia and a group energy of seventy-six guides named Geremyia who now channels the information that I publish in my articles. It explains what happens when we move into the state of unification but continue to hold soul commitments based on Earth's blueprint of polarity. It also provides the rituals needed to remove the conflict so we can be immersed in unification in spite of the state of duality that our society upholds.

Please note that you will undergo a period of adjustment after completing the disavowing of polarity contracts. Apparently, once polarity disappears in our soul matrix, we can finally tap into the twelve levels of awareness and information that the multiple DNA strands provide because we have attained unification. During this time, you are reviewing information from all of your past lives, retrieving what you currently need and discarding what you no longer need. It is almost like a computer program that takes all of the information residing on the hard drive and compresses it by removing spaces to create more streamlined and efficient processing. Needless to say, as you review this massive amount of information, some upheaval is introduced into your life. The less you resist the discomfort by accepting it and moving on, the faster you will complete your compression program. The more you resist it because it feels uncomfortable which is foreign to your experiences by this time, the longer it will take to complete this phase. However, once you have completed this process, you will have access to all twelve levels of information which offers tremendous insight to you as you lead your physical existence. It also creates a state of bliss which is difficult to describe since we are more familiar with the upheaval our increasingly hectic lives seem to bring. Suffice it to say, the blissful feelings are well worth the road one must travel to reach the destination.

Information on Polarity

Earth is a planet that operates on the premise of polarity. You see it everywhere. It exists in nature in the form of tranquil lakes versus violent earthquakes. It exists in the annual cycles in the fruitfulness of summer versus the dormancy of winter. It exists in male and female qualities where the male is characterized as the aggressor or doer while the female is the nurturer or receiver. When your soul decided to enter the polarity library of Earth to learn from its resources, it came with a dual purpose. One objective was to live in the midst of duality and learn to overcome it by discovering Divine Union or unification within this seemingly contradictory framework. Souls are often overzealous in the learning experiences they choose, and discovering unification on a planet of duality probably seemed like a good challenge. Most of you have discovered via your frustration with attaining unification in a polar world that this objective was somewhat ambitious! You have spent multiple lifetimes dealing with conflicts resulting from dual perspectives like good and bad or right and wrong. Finally, you are emerging from this yin/yang pull on your perspective into a mode of acceptance rather than judgment. And, now that you are realizing your objective of balanced living amidst polarity, you are ready to accomplish your second objective which is to represent the principles of Divine Union to other souls through example.

The road toward attaining Divine Union in your own soul path has been fraught with frustration and continues to hold many challenges. This is mostly due to the fact that Earth honors polarity which firmly entrenches the concept in the mass consciousness. You are faced with the dilemma of rising above the mass consciousness which means you must overcome the strong pull on your mental process in order to discover unification. This means you must eliminate certain concepts from your thoughts. For example, there is a best-selling book entitled *Men are from Mars, Women are from Venus.* This book revels in the differences between males and females. Instead of assisting people in discovering methods for bringing male and female energy into balance or unification, it drives a stake between the genders by upholding the differences. The message is to understand and honor the differences in order to live in harmony rather than to bridge the differences by bringing the male and female energy into balance in each individual. It is the balancing of these polar energies that allow couples to join in Divine Union and not the toleration of differences. *Unification will never occur by tolerating differences, even if toleration results in living without conflict. Divine Union only occurs by honoring differences.*

There is another influence beyond the mass consciousness that drives you to separation, whether it is between you and a partner or a friend or a work peer. It exists in the form of your soul contracts or vows. Prior to reincarnating on Earth, your soul agreed to fulfill certain contracts in order to evolve. These contracts were based on the premise of polarity because that is the operating mode of this planet. In order to successfully complete these contracts that were developed among soul groups who honored unification, they had to match the values of the planet on which you incarnated. But, at this point in your evolvement, some of you have moved beyond the concept of polarity, understanding that harmony results from honoring differences. You strive to balance differences through acceptance and love rather than increasing differences through tolerance and separation, upholding the value of unification in your daily existence, be it getting along with your spouse or your neighbor or your boss. Unfortunately, despite your orientation toward union, you are continually being pulled by separation since your soul contracts were based on the principles of polarity rather than unity.

Once you rise above the mass consciousness of polarity and choose the path of unification, your prior soul contracts oppose rather than assist your new reality. You no longer honor the *Men are from Mars, Women are from Venus* mentality because you manifest based on unification. You recognize the common denominators in everyone. There are no divisions. Therefore, you must rescind your prior vows or you will be working against yourself. This can become quite uncomfortable at times because the more you move toward unification, the more you experience the opposite extreme based on the polarity premise of your prior vows. For example, a couple who honors the masculine/feminine balance in both parties versus basing their relationship on the more traditional viewpoint which honors the differences between men and women will still feel conflict. The more each individual moves toward Divine Union by acknowledging similarities and honoring differences, the more separation is attracted because of the violation of prior soul agreements. This can be very confusing since the couple understands and implements unification yet is randomly torn by feelings of separation. It is almost as if they are dealing with a force that is stronger than their desire for unity. Their heart desires to be whole yet their actions defy it. Those who have experienced this division think they are going crazy since they constantly feel like they are working against their desires.

The following ritual was developed to break contracts based on polarity in order to align your soul vows with the principles of unification. This ritual should only be done if you have already minimized polarity thinking and create based on unification rather than separation or if you have completed phase one and two of DNA recoding. Otherwise, you will be breaking vows that can still be accomplished under the premise of polarity thinking which will interfere with your soul path and purpose. *Also, please conduct the following ritual either outdoors on a patio or inside on a good fire-protected surface.* This is a *very powerful ritual,* and some who have conducted it have discovered that the flames from the fire are quite expansive. Sometimes, the flames leap toward the ceiling. Sometimes, they burn or scorch whatever sits directly under the bowl in which the burning takes place.

Ritual for Breaking and Rescinding Polarity Vows

- First, make a copy of the Polarity Ritual on page 233 since you will have to burn it.
- Line a Pyrex bowl or deep pot with aluminum foil and place a blue candle which symbolizes your energy in the bottom. Place the bowl or pot on a fire-protected surface, e.g., a kitchen burner, BBQ grill, trivet, etc. Play music which uplifts you spiritually. Light the candle. You will be burning this piece of paper (contract) in the container at the end of the ritual.
- (IF BREAKING VOWS ALONE:) Sign the following request at the bottom and cut it horizontally in half. Read the first half, then lay it aside. Read the second half. Light the first half with the flame from the blue candle and burn it in the bowl. Repeat with the second half.

- (IF BREAKING VOWS WITH A PARTNER:) Both partners sign the following request at the bottom and cut it horizontally in half. Partner #1 reads the first half while Partner #2 holds the remainder. Partner #1 then takes the remainder and reads it. This is repeated by Partner #2. Partner #1 lights the first half with the blue candle flame and burns it in the bowl. Partner #2 lights the remaining half.
- Let the candle burn completely. Wrap up the waste in the aluminum foil and dispose it.

Polarity Request

- (SAY:) "Divine Creator (3 times). I call on my High Self to envelop me in white Light. I call on my special guides and teachers to surround me with their love and protection. I call on the Planetary Councils that support the movement of Earth away from polarity and toward simplicity and unification to assist me. I ask to be protected and shielded from those who support polarity. I break and rescind all vows and agreements I have made that support the notion of polarity including those that were intended in my higher good and those which were made with the dark forces, as they no longer serve me in my quest for Divine Love. I formally rescind each and every vow and agreement I have ever made in this life or any other life, in both my physical and subtle bodies, and in this time or any other concept of time including those concepts of time that exist beyond my comprehension, that were intended to be implemented under the laws of polarity. This includes both positive and negative contracts since neither will support the concepts of simplicity and unification. These contracts are to be replaced with contracts that support simplicity and unification that will lead to advancement and greater knowledge. Therefore, I break and rescind each and every vow and agreement I have ever made that supports polarity. I ask for my High Self to be surrounded with white Light and that I be provided with new pictures of reality based on the principles of unification and integration. I claim my wholeness, both individually and as a part of the greater whole. I ask my High Self to manifest the highest possibilities of me integrating all I have requested so that the power of that can be fully manifested in my experience. I thank my High Self, my guides and teachers, and the Planetary Councils for assisting me in the healing I have just received. Thank you."

Signature: _____   Signature: _____

Ritual for Fusing Vows of Union

- There is a final step for those who have conducted the rescinding of polarity vows with a partner. In order to be on mission and complete the work you and your life partner have contracted to do on Earth, you must re-fuse or meld your vows based on the premise of unification. Each person cuts a small piece of hair and places it in a bowl. Thoroughly mix the hair together. Then, each partner reads and signs the following vow. Make a copy of the vow since you will need to burn it. Cut the vow in half and retain the portion you have signed. Next, ignite the half of the vow you are holding. Place each half in the bowl with the hair. Be sure the entire contents burn thoroughly, lighting the paper or the hair again if it burns out.

- (SAY:) "Divine Creator (3 times). Archangel Zophkiel from the Realm of Balance (3 times). I call on my High Self to envelop me in white Light. I call on my special guides and teachers to surround me with their love and protection. I call on the Planetary Councils that support the movement of Earth away from polarity and toward simplicity and unification to assist me. I request a re-fusion of my contract with my life partner, _____, based on the principles of unification and integration. I claim my individuality and my wholeness through this union. I am one with the universe. Thank you."

Signature: _____   Signature: _____

About the Author and InterLink

Anne Brewer has studied metaphysics for ten years. Prior to that time, Anne spent fifteen years in corporate America in marketing assignments. For the past five years, she has been President of her own marketing consulting company. Anne's metaphysical client base extends internationally. Her channeled articles frequently appear in such metaphysical publications as *The Sedona Journal of Emergence, Pathfinder,* and *The Edge.* She wrote *Power of Twelve: Achieving 12-Strand DNA Consciousness* in 1996 during her personal experience regaining her twelve strands of DNA. She is currently working on a second book entitled: *Geremyia's Gems, Jewels of Wisdom from the Other Side.* This book includes channeled information from her team of 76 guides who call themselves Geremyia and the Spirit of 76. She has been honored to apprentice with nationally reputed healers and continues to expand her knowledge through ongoing education. Her company, InterLink, is dedicated to eliminating obstacles that are hindering a person's growth and increase spiritual awareness through the following methods:

- Soul-Clearing
  Identify soul-level sources of relationship, health, career, money, etc., problems and break the cycle

- Property Clearing
  Remove negative energy from homes and offices and install permanent protection

- Holographic Repatterning
  Shift from negative to positive behavior patterns in a one-hour session

- Chakra, Subtle Body and Endocrine System Balancing
  Identify the causes of energy blockages and removes them from the physical and energy bodies

- Archangel Realm Readings
  Examine your life purpose as defined by your primary
  Archangel realm and open blockages to all seven realms

- Soul and Spirit Guide Profiles
  Learn your soul origination and the names and
  purposes of your spirit guides

- DNA Recoding
  Acquire twelve versus two DNA strands for full
  consciousness

**To contact Anne Brewer at InterLink for a personal
appointment or to schedule a workshop:**

InterLink
**5252 W. 67th Street
Prairie Village, KS 66208
913-722-5498 *(phone)*
913-722-5497 *(fax)*
www.interlnk.com *(Internet)***